GUIDE TO AS

(Previously entitled *The P*

GUIDE TO ASTRONOMY

(Previously entitled *The Pan Book of Astronomy*)

JAMES MUIRDEN
F.R.A.S.

REVISED AND BROUGHT UP TO DATE

PAN BOOKS LTD : LONDON

First published 1964 by
PAN BOOKS LTD
33 Tothill Street, London, S.W.1

ISBN 0 330 02775 1

2nd (Revised) Printing 1972

Printed in Great Britain by
Richard Clay (The Chaucer Press), Ltd, Bungay, Suffolk

CONTENTS

PART III – AMATEUR ASTRONOMY

LIST OF TABLES

INTRODUCTION

THERE ARE three types of astronomical book. The first is a treatise devoted to some branch of the study, of interest mainly to the serious student. The second is a guide to practical observation for a newcomer to the science. The third is a general introduction containing sufficient information not only to enable the reader to tackle more advanced works, if he feels so inclined, but also to encourage him to make a beginning in the fascinating pursuit of amateur astronomy; and this book is intended to fall into the latter category.

As with all compromises, there are drawbacks; it is physically impossible to squeeze so tremendous a field into three hundred pages, and omissions have to be made. Unfortunately these always seem to be at the expense of astronomy's real battlefields, of which the main one is the nature and origin of the universe, while relatively homely departments such as the Moon and Mars are dealt with in detail. This is misleading, and also makes for complacency; the cut-and-dried departments of the science are easy to deal with, but they are in many ways the least exciting. I have therefore tried to be less partisan, even though it means following several lines of research that end with a large and stubborn question-mark. Astronomers are now so active that some of these riddles may have been resolved by the time the book appears in print, but this is an unavoidable risk!

The major part of this book therefore deals with the province of the professional astronomer; and while it is true that there is still a good deal of scope for the patient amateur, his inquiry must clearly be limited to rather less spectacular fields. I have therefore confined observational matters to a separate section, so that they can be ignored by those wishing to conduct their astronomy solely from the fireside. There are arguments for and against such division, but I hope it meets with general approval.

The seven years that have elapsed since the first edition of this work have seen changes in our astronomical knowledge that necessitated a considerable amount of re-writing. A number of friends and reviewers have pointed out miscellaneous errors in the text, and, as before, I would like to repeat my particular indebtedness to John Larard for supplying information for Chapter 30. John Murray kindly assisted in the preparation of the new Moon map in Fig. 8.

JAMES MUIRDEN

PART ONE

The Solar System

Astronomy falls naturally into two basic departments: that dealing with the stars and the colossal systems or galaxies into which they are massed, and that confined to the relatively tiny bodies orbiting around our own star (the Sun), which form the solar system. Though insignificant on the cosmical scale, the solar system is clearly of the greatest practical importance to us, and this is a good reason for giving it priority.

CHAPTER 1

Man and the Universe

NOW AND then everyone is an astronomer. It may be a glimpse of the evening star, or of a flashing meteor, or a casual glance at an eclipse. Day and night; the course of the seasons; the transition from summer to winter – all these have their roots in astronomy. It is a science which sways our circumstances and, indirectly, our lives.

The primitive peoples lived closer to nature than the civilized societies of today, and it is hardly surprising that astronomy figured so strongly in their routines. Today, but for different reasons, it is once again a dominating science. It has become the vested interest of a rapidly-expanding technology. We no longer gaze at the Moon with feelings of remoteness; when Mars drifts into the telescope's view our emotions are less of wonder than of expectation. Only the stars, it seems, are permanently protected from man's stirrings by the sheer inconceivable gulf of space.

This vast scale of things is one of astronomy's attractions. On the Earth mankind can plough its furrows where it wills; its ability has more or less reached the fearsome stage of compatibility with nature. So under these circumstances it is refreshing to gaze into the night sky and know that nothing can swerve the stars from their inevitable deaths and the planets from their eternal courses around the Sun. We can predict with a certainty that is almost terrifying what the night sky will look like in 100,000 years' time, and if we do not extend our ambition into the millions it is through no lack of confidence in nature, but rather through the inability of astronomers to observe to the right degree of refinement.

At the same time, the very vastness of space acts as a strong discouragement; people complain that they cannot visualize the interstellar or intergalactic scale, and so refuse to investigate the matter further. This is as short-sighted as it is sad. In the first place there is no need to 'visualize' the universe in order to gain some idea of its workings – for no astronomer can really comprehend the enormity of his field of study. It is simply a matter of getting used to dealing with very large units of distance. On the Earth we might arbitrarily define 1 foot as a small distance, 1/1,000th of a millimetre as a very small distance. The astronomical equivalents of 'very small' and 'small' could be 1 mile and 1 light-year (5,880,000,000,000 miles). We can no more imagine 1/1,000th of a millimetre than a million miles – but no one is afraid of looking through a microscope! And at the same time there is no doubt which is the more impressive.

More discouraging is the way the universe has steadily expanded in response to man's probings. With a few exceptions the naked eye can see only a few thousand light-years away from the Earth, yet the ancient Greeks were hardly prepared to deal with such unthinkable gaps in space; why, they asked reasonably, should the gods waste so much room? Until recently, however, we had no concrete evidence for distances in space. The first proper measure of a star's distance was not made until 1838, and even now some of our estimates are probably not entirely satisfactory. There are many difficulties to be overcome and the whole matter makes a fascinating story (Chapter 18).

At the moment, however, our astronomers stand in an enviable position which might well turn into a frustrating one, for astronomy is poised on the brink of at least partial understanding of the greatest problem of all: the nature of the universe. What might be called the Cosmological Controversy reached a spectacular stage in February 1961 when Professor Ryle published his results in favour of an evolving universe, as opposed to the 'steady-state' or static universe of Hoyle and other prominent astronomers. These rival theories are described in full later on; what concerns us for the present is the fact that practical observations are possible at all. Previously all cosmological theories have been theories only, but the recent enormous advances in radio astronomy have brought the matter to a new and possibly decisive stage. We are beginning to sift out definite evidence, and it does not take scientific knowledge to thrill to the way the universe is reluctantly disclosing some of its secrets.

Oddly enough, progress concerning the more homely matter of the origin of the solar system (the Sun and its attendant planets, one of which is the Earth) has been much less spectacular. This is because we lose an ally which favours us in deeper probes into space, and this ally is time. We see everything either by the light it emits (such as a star) or reflects (such as a planet, which reflects the Sun's light). Light travels very fast indeed by terrestrial standards, but is not so frisky on the cosmical scale; at its velocity of 186,000 miles per second it could gird the Earth in $\frac{1}{7}$th of a second, but it takes nearly $8\frac{1}{2}$ minutes to reach us from the Sun, and $5\frac{1}{2}$ hours to reach Pluto, the outermost planet. If we then follow its journey from the Sun to the nearest neighbour star, it will take $4\frac{1}{3}$ *years*. This leads to the expression 'light-year', which is simply the distance light travels in one terrestrial year.

Because of this interval it follows that we see the star not as it is now, but as it was $4\frac{1}{3}$ years ago, and the farther we look into space the more dated our knowledge becomes. The world's greatest telescopes can see objects whose light has taken several thousand million years to reach us. So it follows that if a theory is advanced, we can use it to work out what should have been the state of affairs so many million years ago, and then check up by looking at these regions of space that are so remote both in distance and in time. This is just how Ryle reached his conclusions about the evolution of the universe. But unfortunately the solar system is not so co-operative; we have no history to look up. For compared with the stars even the outermost planet is alarmingly close.

Our absolute isolation in space is brought home best by imagining everything in terms of a small-scale model. Shrinking the Sun to the size of an orange reduces the Earth to a grain of sand circling about 25 feet away. Pluto is a much smaller grain of sand about 300 yards away. But we should have to walk 1,400 miles before finding the nearest star – another orange. It is clear that on the stellar scale the solar system is an extremely compact unit, so compact as to be utterly insignificant. This blow to our pride is but one of the many we have received since the discovery of the telescope.

The Sun is a star – a quite ordinary star – and it is just one of perhaps 100,000,000,000 stars that collectively make up the local system or galaxy, usually referred to simply as the Galaxy. Galaxies are very common in space, for they are the units of the universe in the same way as atoms are the units of matter. Wherever we look we see galaxies, and the number detectable with the largest telescopes runs into the thousand million.

All the individual stars visible in the night sky belong to the Galaxy, for the other galaxies, even though they contain millions of stars, are so distant that they appear merely as dim blurs of light. Over the whole sky the naked eye can see perhaps 6,000 stars, while a large telescope will count several million. This is considerably less than the population estimate because certain vast tracts of interstellar space are filled with tenuous obscuring matter – dust or gas, or both – which blocks out the light from more distant regions. In some ways this is fortunate, for recent studies with radio telescopes have suggested that some parts, especially near the centre of the Galaxy, would light up the sky more effectively than the Full Moon!

The gap of $4\frac{1}{3}$ light-years between the Sun and its neighbour is a reasonable average of interstellar distances, and it turns out that the Galaxy's population is grouped in a colossal spiral system about 80,000 light-years across. At the centre is a relatively dense nucleus with a diameter of perhaps 20,000 light-years, and from this trail the immense spiral arms. These arms are slowly rotating, like some ponderous catherine wheel; in the region of the Sun, far out on one of the arms, it takes over 200,000,000 years to achieve one revolution. In addition to this general spin all the stars have random motions of their own, but they are so far apart that the likelihood of the Sun, for instance, colliding with or even passing near another star is vanishingly small. For on the scale model orange-sized stars are separated by well over a thousand miles.

Nevertheless the possibility of just such an approach was raised in the first year of the twentieth century by two American astronomers, Chamberlin and Moulton, in their endeavour to explain the origin of the solar system. This question, the birth of the planets, is obviously of tremendous significance; in some ways it concerns us even more deeply than the far vaster issue of the nature of the universe. Were the planets torn from the Sun as a great mass of gas? If not, where did their substance come from? More far-reaching: are planetary systems common throughout the Galaxy? If so, the scales are heavily weighted in favour of there being millions of Earth-like planets with their precious cargo of intelligent civilization. Life so widespread cannot be an accident. Could it be, in fact, the natural climax to the physical processes responsible for the workings of the universe?

As soon as it was realized that the Sun is but one star among many, astronomers and philosophers tried to account for the formation of the planets as something intimately concerned with the Sun's history. The first widely-known suggestion was the famous nebular hypothesis of Laplace (1796), and his basic reasons summed up the grounds of most later theories. Before mentioning these, however, it will be as well to draw an outline picture of the solar system itself.

At the centre is the Sun, our star, an 864,000-mile globe of luminous gas which pours radiation into the surrounding space at an inconceivable rate: every second it discharges as much energy as would be released by a thousand of the largest hydrogen bombs we are ever likely to construct; and it has been keeping up this output for millions of years. It spins on its axis once in 25 of our days, and around it, in the plane of its equator, revolve the nine major planets. Their distances range from 36,000,000 miles in the case of Mercury to the 3,600,000,000 miles that separate Pluto from the central furnace. The Earth, at a distance of 92,900,000 miles, comes third in the sequence, and its temperature comes roughly midway between the fierce extremes of heat and cold suffered by the innermost and outermost planets.

These planets, spinning on their axes in various times, revolve around the Sun in roughly the same plane and in the same direction. Their actual periods of revolution, or years, vary according to distance. Mercury moves very fast around a relatively short orbit and takes only 88 days to circuit the Sun, whereas Pluto takes 248 years because it is ambling along a much longer path. All the planets except Mercury, Venus, and Pluto have one or more satellites, and these planet-satellite systems form rough miniatures of the solar system itself. It is obviously a very orderly family, as Laplace was quick to point out; if we travel to a point in space at right-angles to the plane of the system we shall find that the rotation of the Sun, the rotation of the planets, and the direction of their orbital motion is all in the same sense.[1] Obviously there has been some great co-ordinating factor, and he decided that the Sun, planets, and satellites had all been formed at roughly the same epoch, out of the same vast cloud of primordial gas and dust.

Laplace pictured this disk-like cloud of matter slowly rotating and cooling as it rotated, so that it began to contract. For simple physical reasons the rotation speeded up. Now just as someone can remain standing on a rotating wheel until it reaches a certain critical velocity, when it flings him off to the side, so the outer reaches of the gas-cloud became detached in great rings as the nebula spun faster. Each ring from then on

[1] Since Laplace's time 6 satellites have been discovered which revolve around their primaries in the wrong sense, and therefore offer certain objections to his basic contention.

led a life of its own, and after nine or possibly ten had been formed the central mass finally contracted into the fiercely hot body that we now see as the Sun. The solid particles in each ring gravitated towards each other and eventually formed, though on a much smaller scale, a similar rotating nebula, which in turn contracted and threw off rings. The central masses, however, were much too small to evolve into shining stars; instead they quickly cooled and solidified, and we see them today as the planets with their attendant satellites. The tenth ring is required to account for the crowd of tiny minor planets or asteroids which circle between the orbits of Mars and Jupiter, and which could be the legacy of a planet that failed to grow into a proper body.

Laplace's theory has had to be abandoned, partly because of discoveries made since it was advanced but mainly on purely mathematical grounds. If we consider a body revolving around a point we can arrive at a value for its energy known as its angular momentum, and the greater its distance from the point (other things being equal) the greater the momentum. The central point of the solar system lies inside the Sun's surface, so that despite its colossal mass its angular momentum is relatively small. On the other hand the planets are insignificant compared with the Sun, but each one has a large angular momentum because of its great distance from the central point. On the nebular hypothesis, for the condensation processes to be possible, the angular momentum should be concentrated in the Sun. As things are, the most that could have happened would be the formation of very tiny planets; the large globes that we see today could not possibly have come into being.

This destruction of the nebular hypothesis led to a vacuum that was not filled until 1900, when Chamberlin and Moulton attacked the problem from an entirely different angle – though with more or less the same end-product. They invoked a 'rogue' star which passed very close to the Sun, at a distance of just a few thousand miles (of course we can, if we like, consider the Sun itself as the rogue!). The result would be an awesome struggle as during the critical hours the outer layers of the stars were disrupted and scattered throughout the neighbouring space as a great nebula, part of which was dragged away by the intruder while the rest was left slowly circling the Sun. Once again accretion processes took place and finally built up the planetary system that we see today. The angular momentum objection is overcome because there is no need to introduce a uniform disk of matter which subsequently split up. The planets were formed directly from the great gouts of material that spurted from the Sun's tide-rent surface.

Other theorists followed this line, among them Sir James Jeans, who suggested that the passing star drew from the Sun an immense cigar-shaped filament of matter which immediately condensed into the planets. His point was that just as the widest part of the 'cigar' occurred near the middle of the filament, so the largest planets are roughly midway in distance from the Sun. In other quarters an actual stellar collision was mooted. But it is now known that no mutual encounter could explain why the solar system is so extensive; if the planetary material were really dragged from the Sun, the total extent of the system should be bounded by Mercury's orbit. There is obviously something very wrong here!

The great drawback of these 'catastrophic' theories, which depend for their success on what amounts to a stellar disaster, is that such disasters are extremely unlikely. Stars do not have to pay a very high premium against assault, since the chance of their colliding with a neighbour is something like once in a million million years – many times the age of the Galaxy! It is true that these tidal theories do not demand an actual collision, but even so the probability is a remote one. If all solar systems had come about in the same way, ours might well be the only one in the Galaxy. This is elevating for our ego, but unsatisfactory on more objective grounds – especially since there is evidence of other planetary systems besides our own.

It therefore seems that all catastrophic theories must be treated with a certain amount of reserve. The nebular hypothesis avoided the difficulty, since all stars presumably condensed from gas-clouds and might well be expected to have produced similar systems. Indeed, planetary systems would be very much the rule rather than the exception. Is there any other way in which we can retain this rather necessary criterion?

Hoyle has put forward an idea which avoids this trap, although few astronomers support it. He suggests that the primordial matter came not from the Sun itself but from a nearby star that exploded. Such exploding stars, or supernovae, appear from time to time among the multitudes that throng the Galaxy; one is visible with the naked eye every 300 years or so, which means that they must be fairly common. When this disaster occurred, some 5,000,000,000 years ago, its material was blasted throughout the local regions of space with quite catastrophic violence. The Sun, together with many other nearby stars, came under this celestial barrage and was left smeared with an unwelcome cloud of debris that slowly settled into a disk and condensed into solid particles and subsequently into planet-sized bodies. The fact that there are no longer any stars very near the Sun is no objection, for it is widely believed that stars are formed in great clusters which gradually break up, and at that remote epoch the Sun itself was a very young star. In fact it is probably hardly superior at all to the Earth in age.

More widely accepted are the theories resulting from the suggestion of Carl von Weizsäcker, a German physicist, that the Sun gained its primordial cloud in a very unspectacular manner: by drifting through one of the many immense clouds of interstellar dust, or nebulae, which throng the Galaxy. If the Sun had passed through such a nebula early in its history the stage would be set for subsequent processes; more than that, the same thing must have happened to innumerable other stars as well. This is pleasing news to those who like to think of life as a common product of the universe, and at the moment von Weizsäcker's basic idea has met with general acceptance; the main controversy now is over the actual process of aggregation, and this involves highly technical wrangling.

With this conclusion we reach the present state of man's inquiry into the part of the universe immediately around him;

his wider probes must be saved for a later chapter. The ultimate problem – of how life on the Earth began – can hardly at this stage be called an astronomical question. For instance, we do not yet know if terrestrial life is the only possible form of life. The astronomer observes greenish patches on Mars which are evidently living, and the biologist assumes that they consist of a life-form known on the Earth, since otherwise we cannot investigate the matter further until a manned spaceship lands there. But on the question of whether or not life is all built on the same basic pattern, and has ultimately the same destiny, the universe has so far remained silent.

CHAPTER 2

The Sun

WE OFTEN speak fondly of Mother Earth without realizing that the globe we live on is anything but motherly. It is lightless and heatless; most of its surface is covered with water, and much of the rest is mountain, desert, or forest. Only an alarmingly small proportion is really hospitable, and without the Sun's gentle heat any sort of life at all would be completely out of the question. Ultimate thanks must go not to the Earth but to the star around which it revolves.

For the Sun is an ordinary star, if we can speak of any star as being ordinary; Chapter 18 explains how there are different stellar classes, and it is therefore rather like speaking of an 'ordinary' car. We can at least say that it is unremarkable among its 100,000,000,000 fellows. If anything its size and mass are both rather below average, but its light-output is correspondingly higher.

Because it is so unremarkable, the Sun's value as a prototype is obvious. We are so close to it that it can be studied in considerable detail (no other star, not even the nearest, shows a disk in the largest telescopes), and up to a point we can infer that what goes on in the Sun goes on in other stars as well. If galaxies are the molecules of the universe, stars are the atoms; and the Sun holds the key to the understanding of those fundamental units from which our cosmological knowledge must be built. It is therefore small wonder that since the late nineteenth century the Sun has been the subject of most intensive study by observatories all over the world. It brings together many specialized branches of science, from nuclear physics to meteorology, and solar observation is one of the most important fields of astronomical research.

The birth of the Sun is so intimately concerned with other Galactic matters that it is best left for a later chapter; it is sufficient to mention here that it probably condensed from a vast cloud of gaseous matter some 6,000,000,000 years ago. The problem of how it has managed to maintain such a stupendous output of energy for at least a thousand million years (the duration of life upon the Earth) has puzzled physicists for a long time. One idea, following directly from its condensation from the nebula, was that the Sun is all the time cooling and therefore contracting, the contraction in its turn raising the temperature by a compensating amount. Every cyclist knows, when pumping up a tyre, that compression produces heat, and it was calculated that an annual shrinkage of about 400 feet (which would of course be imperceptible even over hundreds of years) would keep the Sun shining. However, there is an obvious drawback to the theory: the Sun cannot contract indefinitely, and on this principle its total life-time could not be more than about 25,000,000 years.

To get over this difficulty various regenerative processes were suggested, in which the same material was used again and again; but this was clearly a case of using sandbags against the ocean. Another suggestion, put forward at the same time (1853), attributed the heat to the impact of tiny interplanetary particles that were drawn towards the Sun's surface. This means that instead of shrinking, the Sun is all the time becoming larger and more massive. This in turn would have a clear effect upon planetary motions, making their years steadily decrease, and needless to say this speeding-up has not been observed.

It is easy to be amused at these rather clumsy artifices, but before the advent of nuclear physics there seemed to be no way in which energy could be produced other than by friction or combustion – which meant that the Sun had to be literally burning its substance in the manner of a coal fire. No wonder it needed rather a lot of stoking! The true answer – or what we believe to be the true answer – could only emerge when the immense energy-reserves of the atom had been discovered. We now know that the Sun, which is composed principally of hydrogen, produces its colossal quota of radiation by converting its hydrogen into helium.

All matter consists fundamentally of some of the ninety-two

natural elements known to science, and these elements take their place in an orderly sequence in the Periodic Table. What is more, the atoms of these elements all consist of the same basic particles. It therefore follows that by interfering with its structure, an atom can be changed into the atom of another nearby element. This process is known as transmutation, and it takes us back to the days of the alchemists and their struggle to make gold from lead. Even though their methods were questionable, their quest was at least theoretically justified.

Transmutation has only recently been accomplished in the laboratory, and even then only among the heavy, unstable elements such as uranium, which require a relatively low temperature. The Sun, however, is an atomic furnace. Its surface temperature is 6,000° C, but near the core it climbs to an estimated 20,000,000° C. Under these conditions hydrogen atoms are stripped and re-formed into helium atoms, producing sufficient energy to maintain the process until the supply of hydrogen gives out.

The precise way in which transmutation works is extremely complicated, and in any case is not perfectly understood; there also appear to be two basic processes. But one thing is certain: a slight amount of mass is lost in the process. The nuclei of a hydrogen atom are slightly too massive to form the nucleus of a helium atom; they therefore lose this mass in the form of energy. In the hydrogen–helium reaction less than 1 per cent of the atomic mass is lost in this way, but even so the process is carried out on so tremendous a scale that this is ample to maintain the Sun's radiative properties. Not only that: there is sufficient fuel to last the process for many aeons yet. It is a sobering fact to realize that although the Sun is losing mass at the rate of nearly 5,000,000 tons per second, it has sufficient hydrogen reserves to keep going for many thousands of millions of years.

In fact, if we can believe modern theories, the Sun is doing its job rather too well: it is getting hotter. The rise of temperature is negligible on the human time-scale, but by about AD 10,000,000,000 it will be so hot that the Earth's oceans will be literally boiling off the turgid, plastic crust, while even the outer planets will wither in the searing heat. But its fury will not last long. Exhausted of hydrogen, its atomic furnace will die down, and in just a few million years it will collapse into a cool globe about the size of the planet Jupiter. After that only its dark, stricken planets will bear witness to its past glory.

There is not much chance of human eyes witnessing the Sun's ultimate downfall, but even in its present, relatively quiescent state, it shows many features of interest. Its apparent surface, the surface we see with the naked eye or through a telescope, is called the photosphere, and this is where sunspots occur. Above the photosphere is the solar atmosphere, which is relatively so faint that we cannot normally see it except when the photosphere is blotted out during an eclipse.

With a low-power telescope the photosphere itself appears utterly smooth and featureless, but this is really far from the case; a large telescope working at a high magnification will reveal it as a mass of tiny bright specks, known as granules. They are 'tiny' only in the solar sense, of course; a granule may measure anything up to 1,000 miles across its longest diameter, and their roughly elliptical shape gives rise to an alternative and more descriptive name: 'rice-grains'.

The solar granulation was first observed about a century ago, in 1862, by James Nasmyth, an English amateur astronomer who was also the inventor of the steam-hammer. The theory put forward then, and held until quite recently, was that the granules represented the uprush of hot streams of matter from the interior, while the spaces in between, which are darker and therefore presumably at a lower temperature, are the currents descending again into the interior. This situation is like a very mild case of boiling (with the difference that there are no actual bubbles of gas involved), and the fleeting appearance of each granule, which has a lifetime of about 3 minutes, lent support to this idea.

However, astronomers have been restless. Observation of any fine detail, whether it be on the Sun or on a planet, is hampered by the Earth's atmosphere. It may come as a surprise to know that the primary requisite for this type of observation is not so much a clear sky as a steady atmosphere. Normally the upper air, at an altitude of about 50,000 feet, is a boiling mass of currents at different temperatures; these currents refract the light by different amounts and may make the image so turbulent that the details are lost in the general confusion. Steadiness of 'seeing', as it is called, is at least as important as sheer telescopic power. This is borne out by the fact that some of the first photographs of the solar granulation, taken about 80 years ago and with extraordinarily primitive apparatus by the French astronomer Janssen, are almost as good as the best produced by observatories today.

Transporting telescopes to mountain-tops has its draw-backs, and in any case the highest mountain is still well below the main level of interference. The only real solution is to launch apparatus to the borders of true space, either by rocket or by balloon.

Satellite-carried instruments are still a thing of the future, but balloons are far less ambitious, and in 1957 and 1959 some very interesting experiments were carried out by the astronomy department of Princeton University, under the direction of Dr Martin Schwarzschild, called Project Stratoscope. A 12-inch aperture telescope, which is small by professional standards, was carried to a height of 80,000 feet by their balloon *Stratoscope I*. It was then automatically pointed at the Sun, after which a camera came into operation and took a large number of pictures on 35-mm. film; finally the entire apparatus returned to the Earth's surface. The results of their endeavours are the finest photographs yet obtained of the solar granulation, and recently they have succeeded in launching a new balloon, carrying a larger telescope, with the object mainly of photographing the planets. It seems that the Princeton workers are opening a new era in astronomical observation.

The photographs, of which only a few were perfect, show the granules as sharply-defined as in a mosaic pattern. This suggests that the photosphere is even less turbulent than has been believed; the 'boiling' theory is inadmissible, and the nature of the granulation still has to be satisfactorily explained. One suggestion is that it is similar to the pattern seen on the surface of a thin layer of paraffin wax when it is gently heated.

Every now and then a minute dark spot appears among the granules, growing in a matter of hours to what is known as a pore. This is the beginning of a sunspot.

Sunspots provided the world with some of its earliest astronomical observations; the enlightened Chinese periodically recorded 'birds flying in the Sun' when it shone through thin cloud and could be looked at safely. In Europe, however, such phenomena were discounted by religious feelings. Even when Galileo turned his new telescope to the Sun and not only saw spots, but also watched them move across the disk as it rotated, the Church censored his observations. It was not possible for so divine a body to be imperfect.

If the Sun really were perfectly featureless solar astronomy would be a very dull business indeed. Luckily hardly a day passes without at least a couple of small spots being visible, and it is fascinating to watch their birth, development, and decay.

Sunspots appear dark not through pigmentation but because of their lower temperature. They are about 1,000° C cooler than the surrounding photosphere, and this reduces their visible radiation – though if they could be seen by themselves they would shine brilliantly. Their basic composition is very similar to that of the photosphere, which consists principally of hydrogen, helium, and calcium but altogether contains about seventy elements. However their relative coolness has allowed certain compounds to form which would be immediately decomposed outside the shelter of the sunspot.

It is worth pausing here to consider how it is possible to analyse the composition of an inaccessible object such as the Sun, or, worse still, a star. This branch of astronomy, known as spectroscopy, is a vital one. Without the spectroscope our knowledge of the universe would have hardly advanced at all during the last century, and it is quite unfair to attribute the progress of astronomy to the building of larger and larger telescopes. This is only half the story; the spectroscope provides the other half.

Light is a form of radiation known as electromagnetic radiation, which itself encompasses all types of emission from X-rays at one extreme to the Third Programme at the other. The difference is one of wavelength, and we can compare this type of radiation, albeit rather inaccurately, with sound-waves. On a piano top C sounds much higher than bottom C, but it is essentially the same phenomenon; it simply has a much shorter wavelength. In electromagnetic radiation X-rays correspond to top C, radio waves to bottom C, while visible radiation occurs near the middle of the keyboard. The whole band can be encompassed by one extensive spectrum (Fig. 1).

Radio waves are usually measured in terms of metres, but when we reach the visual spectrum the wavelength has become so short that it is necessary to introduce a new unit: the Ångström (Å), in which 1 Å equals 1/100,000,000 of a centimetre. The band of visual radiation runs from about 4,000 Å (violet) to 7,000 Å (red) – the well-known spectrum whose tints, when combined, produce the effect of 'white' light. Radiation shorter than 4,000 Å is invisible, and in fact most of it is absorbed by the Earth's atmosphere and so cannot be detected in celestial objects; it is known as ultra-violet light. Beyond the red end of the spectrum comes the invisible infra-red, which merges into heat waves and finally radio waves, which can have wavelengths of thousands of metres.

The job of the spectroscope is to sort out the light passing through it into its component wavelengths, and this it does either by using a prism or a very finely-ruled plate known as a diffraction grating. The net result is a long band of colour familiar to everybody, and we get it when we point a spectroscope at the Sun. It is known as a continuous spectrum, for obvious reasons.

This by itself is not very informative, but closer examination shows that crossing the band are thousands and thousands of very fine dark lines. This means that at these particular wavelengths we are receiving low emission from the Sun, and this in turn is due to the different elements present. In other words, each of these lines represents an element, and they can be identified by comparing the solar spectrum with sample

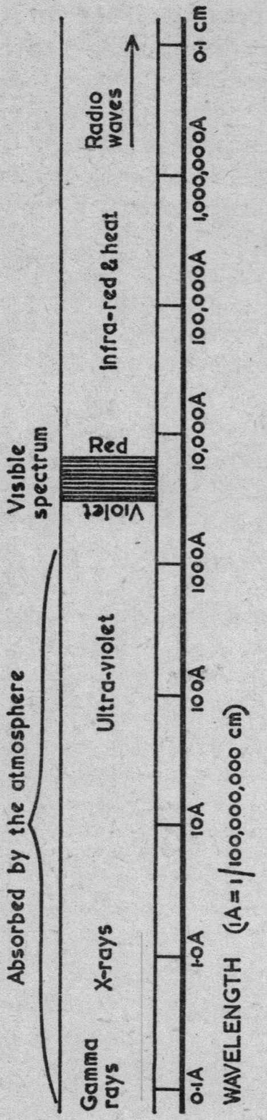

FIG. 1. *The electromagnetic spectrum.* This diagram shows only one extreme of the spectrum, for radio waves can be many thousands of metres long. Much of the radio band suffers atmospheric blockage similar to that for ultra-violet light, and in general radio astronomers can receive cosmic emission only in the range between 0·5 centimetres and 10 metres.

spectra of individual elements. Most elements produce dozens of lines in the visible spectrum – iron produces hundreds – and it is a straightforward matter to check their coincidence.

We must be careful when talking about a substance being 'in' the Sun, for there are two ways in which it can reveal itself. If we take a gas and heat it until it begins to glow, it will give a spectrum showing a number of bright lines against a dark background; something that is known as an emission spectrum. If we then cool the gas and place a light source behind it, we will obtain a continuous spectrum (due to the light source) crossed with dark lines (due to the gas). This is known as an absorption spectrum, since in this case the element has absorbed light of certain wavelengths. In both cases the lines are in precisely the same positions, but they appear bright or dark according to whether the element is emitting light or absorbing it. It is evident that the Sun gives an absorption spectrum, which is due to the elements in its atmosphere rather than in the photosphere – although there is no reason to suppose that their composition is markedly different. It is also important to remember that the terrestrial atmosphere imprints its own lines on the solar spectrum, and these must be carefully weeded out.

Spectroscopy turns up again and again throughout astronomy, but it is time to return to sunspots, which begin their lives as tiny pores, or groups of pores, on the photosphere. Many pores die away almost as soon as they form, but others grow rapidly, diffusing into other nearby companions until a proper spot is formed. No spot much smaller than the Earth's diameter is really worthy of the name, and some groups have achieved an overall length exceeding 100,000 miles. The largest ever seen, in April 1947, had an area of over 7,000,000,000 square miles.

What usually happens is that two spots form close together in roughly the same latitude, so that due to the Sun's rotation one appears to lead the other across the disk. They are linked together by a mass of outlying pores and smaller spots, and after maximum development the following spot dwindles, the outliers decay, and finally the solitary leader shrinks and disappears. Sometimes lone spots develop, and occasionally a group appears which is a mass of irregular spots. They are known as unipolar, bipolar, or multipolar according to the number of main nuclei.

A sunspot itself consists of two distinct areas: the dark central region, or umbra, and a surrounding annulus midway in intensity between the umbra and the photosphere, known as the penumbra. Once again the two tones are the result of temperature differences, the umbra being considerably cooler than the penumbra, but it is a mystery why the division should be so sharp. Just as mysterious is the range of their lifetimes. The average spot-group lasts a week or so, but some have lasted for many months, while one observed in 1840–1 is said to have lasted for a year and a half.

We speak of sunspots as being cooler regions of the photosphere, but there is more to the matter than that. One curiosity is that they are often slightly depressed, like very shallow lunar craters, a fact first brought to notice in 1774 by Alexander Wilson, professor of astronomy at Glasgow University. In November 1769 he noticed that as the solar rotation carried a large spot towards its western edge or 'limb', the penumbra nearest the centre of the disk contracted. The Sun appears to rotate in 27 days, since the Earth's orbital motion effectively slows down its true spin, and when the spot reappeared a fortnight later the opposite part of the penumbra, which was now also nearest the centre of the disk, appeared foreshortened. Wilson confirmed this effect with other spots, and it is now known as the Wilson Effect. It can be simulated, in an exaggerated form, by viewing a saucer from an increasingly oblique angle (Fig. 2). Not all spots show the Effect, but many do, and the depression of the umbra below the level of the photosphere usually comes to about 400 miles.[1]

Another property, this time a universal one, has been revealed by the spectroscope. For a long time it has been known

[1] It should be pointed out that while this phenomenon occurs with a fair proportion of sunspots, astronomers are by no means unanimous about its true significance. It may possibly be due to some cause other than depth.

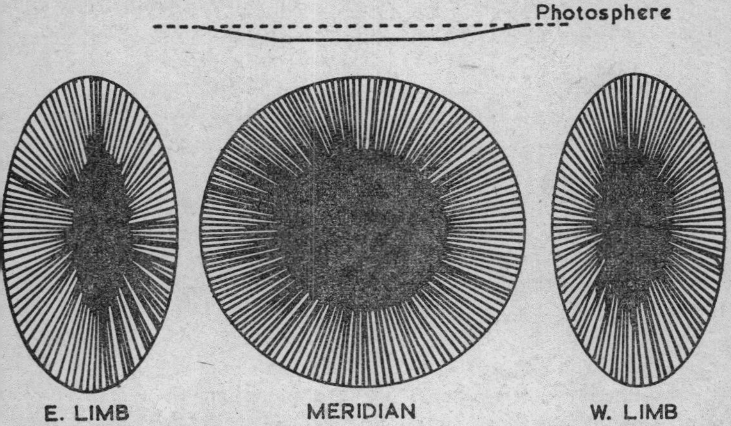

FIG. 2. *The Wilson Effect*. Some sunspots, but by no means all, show a slight displacement of the umbra when seen in inclined view near the Sun's limb. This suggests that they are concave in form, as shown in the profile above. However, many spots refuse to show any shift at all, while some appear to exhibit an opposite displacement – the inference being that they are convex.

that if the light from a chemical substance has to pass through an intense magnetic field, its spectral lines tend to subdivide and split into clusters; the more intense the interference, the greater the subdivision. In the early years of the present century, when George Ellery Hale and his intrepid band of pioneers were struggling to establish the observatory on the top of Mount Wilson in California, they studied the spectra of sunspots. To their amazement some of the lines showed this doubling, which could only mean that sunspots produce their own magnetism. Today this has been confirmed beyond all doubt, although it is rather uncertain whether the spots produce the fields or the fields produce the spots! In short, we still do not know just how sunspots are formed.

Simple telescopic observation over a number of years will show that sunspot activity varies very considerably. Sometimes the disk will show three or four prominent groups, while at others it will be almost featureless for days on end. Oddly enough, this rhythm escaped detection by early observers; the material was there, had anybody thought to analyse it, but as events turned out it was left to an otherwise obscure German apothecary named Schwabe to make the discovery. Using a tiny 2-inch aperture telescope he patiently observed the Sun on every clear day from 1826 until 1850, drawing whatever spots were visible, and as a result of this immensely persevering work he announced a cycle of activity of roughly 10 years. The actual value is closer to 11, but it is by no means

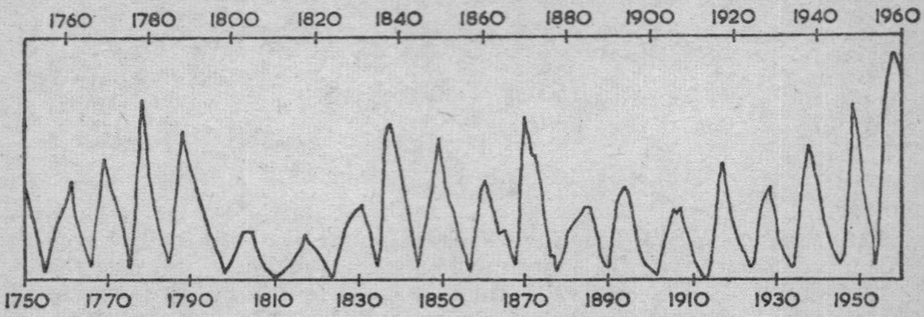

FIG. 3. *Solar activity since 1750.* It is evident, from this graph, that the period of sunspot activity varies considerably; 17 years elapsed between the maxima of 1787 and 1804, but only 7 between those of 1830 and 1837. Moreover, maximum intensity also varies between wide limits. The reason for these secular irregularities is unknown.

regular and is certainly not predictable. Some periods (from maximum to maximum) have been as short as 8 years, while there was a gap of over 16 years between those of 1787 and 1804. One feature which does seem to be consistent, however, is that the rise to maximum is swifter than the subsequent fall. The latest maximum is just past, while the previous one, which occurred in the winter of 1957, was the most active ever recorded.

We have no clue as to why sunspot intensity should fluctuate in this manner, and the very looseness of the period poses its own problems. Moreover, there seems to be a super-period imposed on the 11-year cycle, as Fig. 3 shows: a period of about 90 years linking the strongest maxima with the weakest. If we may forecast based on the evidence, it seems likely that the next cycles in this century will be relatively weak.

Spots not only change in intensity during a cycle; they also shift their general position. On the whole, sunspot occurrence is limited to the region between 5° and 40° N and S of the solar equator; spots are never seen exactly on the equator or in very high latitudes. The first spots of a new cycle are concentrated in the region of 25° to 30°. At first they are small and scattered, but as the cycle progresses they become more numerous and increase in size. At the same time they descend in latitude, so that by the time maximum activity is reached the spots are concentrated in two regions on either side of the equator. In the dying years of the cycle these belts narrow, and at the same time the high-latitude forerunners of the next cycle make their appearance, so that at the time of minimum there are actually two distinguishable zones. This behaviour is called Spörer's Law, and like many other sunspot phenomena remains unexplained.

Sunspots are nearly always associated with 'faculae', bright clouds of luminous gas, usually of irregular shape, that float at altitudes of a few miles above the photosphere. The relationship is not clear, but if a group of faculae are seen a sunspot is quite likely to form in the region before very long; they are the vultures of the Sun. They must not be confused with flares, which are vast outbursts of luminous hydrogen vapour, usually lasting only a few minutes, occurring over sunspots. In addition to being intensely bright, flares emit very short-wave radiation which sometimes has a serious effect on terrestrial communications. This effect occurs through disruption of the ionosphere, which is a shell of electrically charged particles roughly 60 miles above the Earth's surface, which reflects radio emission of certain wavelengths like a mirror. Radio waves, like light rays, travel in a straight line, and normally it would be impossible for two stations separated by more than a few miles to communicate with each other, since the curvature of the Earth's surface would obstruct the waves. What is done, therefore, is to reflect the transmissions off the ionosphere and back to the ground again, and by this means stations thousands of miles apart can keep in contact.

When a flare occurs and the surge of ultra-violet radiation

reaches the ionosphere, its effect is to temporarily neutralize the charged particles; effectively, it destroys part of the radio mirror. In this case the transmitted waves simply soar straight out into space, the receiving station hears nothing at all, and the condition is known as a fadeout. During times of sunspot maxima, when flares are most frequent, there is sometimes almost continuous radio interference for days on end, and occasionally even public short-wave transmissions are affected.

Even more spectacular is the effect of other radiation, consisting of atomic particles and known as corpuscular radiation. This influences the very high-altitude regions of the atmosphere around the north and south poles in a manner described in Chapter 16, producing the shimmering lights we call aurorae. Since aurorae are closely tied up with the Earth's magnetic field, it is not surprising that the emission often has a noticeable effect upon delicate compasses. These 'magnetic storms' always accompany brilliant displays of aurorae.

It is interesting to note that corpuscular radiation travels much more slowly than its ultra-violet counterpart, which of course travels at the velocity of light. Therefore, if a flare is seen, its accompanying fadeout (if it occurs) will be observed at precisely the same time, while the geomagnetic interference has to wait until the corpuscular radiation reaches the Earth. This takes from $1\frac{1}{2}$ to 2 days, so that if a large spot is seen to cross the Sun's meridian, there is at least a breathing-space before the magnetic field is affected. Radio workers are not so fortunate.

Flares always form and remain close to the photosphere – they develop horizontally rather than vertically – but prominences are much more active. They are in fact visible with the naked eye during a total eclipse, when those appearing at the limb of the Sun can be seen projecting into space rather like rosy flames. They occur all over the disk, and are independent of sunspots, but because they are usually less luminous than flares they are only well visible when seen at the limb, in profile.

Flares are relatively small, usually not more than a few thousand miles across, but prominences can reach a colossal size; some have been observed stretching several hundred thousand miles above the photosphere! Like flares, they are eruptions of incandescent gas. But they are much more sedate, and some may last for months before they either collapse back into the photosphere or simply fade away. They are also affected by the solar cycle, but in a rather curious way. They occur in two main belts, one near the equator and the other up in the polar regions (if we can speak of a polar region on the Sun!), well away from the sunspot zone. Those forming in the low-latitude belt more or less follow the cycle, but the polar prominences are most active around the time of sunspot minimum.

There are two classes of prominence: quiescent and eruptive. Quiescent prominences generally take the form of colossal arches of glowing gas, and they usually last for at least one solar rotation. Eruptive prominences, as their name suggests, are relatively short-lived. They are usually smaller than the quiescent kind, but much of the material composing them is ejected from the photosphere at colossal speeds – sometimes as much as 400 miles per second. This means that the Sun's gravitational pull is too weak to drag them back; they have exceeded the 'escape velocity' and soar away free into space. The Sun's escape velocity is 380 miles per second, the Earth's only 7, so space-travellers can count themselves fortunate that they do not live on a globe as massive as the Sun!

Special instruments can detect prominences when they appear fully on the disk, but the average amateur has to rely on total solar eclipses. These, too, furnish the only chance of seeing the Sun's atmosphere with the unaided eye. Total eclipses, which occur when the Moon passes centrally across the Sun (page 58), are not particularly rare – there are often at least two every year – but they are visible from only an extremely limited part of the Earth's surface, which usually turns out to be the Antarctic or the middle of the Pacific Ocean.[1] What is more, they can never last for more than

[1] This is because the Moon is only just large enough to cover the Sun, and a displacement of a few dozen miles from the critical 'central line' means that the observer sees only a partial eclipse.

$7\frac{1}{2}$ minutes, and are usually over much more quickly, so that the precious moments of totality have to be put to the utmost possible use. The last English eclipse was in 1927, and the next is not due until 1999.

When the Moon's shadow sweeps across the landscape and the sky goes dark, the brilliant aureole around the black Moon is seen to consist of two parts. There is a narrow inner ring of rosy light, from which the prominences project, and a far more extensive outer region, pearly white in colour, which may in parts be traceable for a degree or more from the Moon's limb. The inner region is the chromosphere, which gives rise to the absorption lines in the solar spectrum, while the much paler envelope is the corona.

The chromosphere is about 8,000 miles deep, and is clearly the densest part of the atmosphere – though it may come as a surprise to learn that this density is only about 1/10,000 of that of ordinary air. Nevertheless it is very much more substantial than the corona, which unlike the chromosphere has no light of its own: it shines by reflected sunlight. It consists mainly of atomic particles which are streaming away from the Sun due to the sheer force of its radiation, and these particles can be detected in the vicinity of the Earth (the solar wind), so that in a sense we are ourselves involved in the outer reaches of the corona. Needless to say, its density is quite incredibly low.

The particles forming the corona are strongly influenced by magnetic fields, and it was because of this that the Sun's magnetic field was discovered. At sunspot maximum it appears to be distributed fairly evenly in all latitudes, but near minimum the main extension is near the equator, while the polar corona is confined to a few plumes or 'brushes'. These brushes are reminiscent of the lines of force at the poles of a magnet, and in fact the Sun is a gigantic magnet – though why its magnetism should vary with the sunspot cycle remains another unanswered mystery.

The corona may be faint visually (during an eclipse the inner part shines about as brightly as the Full Moon), but it is a very powerful emitter of radio waves, and to a radio telescope it is indistinguishable from the disk itself. In other words the 'radio Sun' is much more extensive than the Sun we see with the eye. This offers a leader to many startling comparisons between optical and radio astronomy; comparisons which will become more and more important when in a later part of this book we investigate the Sun's true place in the universe.

CHAPTER 3

The Moon

THE MOON is a quarter of a million miles away, and this, baldly, sums up its importance. It is a dead, dry world, airless and utterly inhospitable; it is the very antithesis of our home planet. By midday its ashy surface is the temperature of boiling water, while at night it freezes in the deathly chill of interplanetary space. The only colour in its landscapes is grey, and overhead the sky is always black. Only because it looms large and close does it figure so prominently in the fast-emerging era of space travel.

Of the nine major planets in the solar system, all except Mercury, Venus, and Pluto have satellites. Some of these are larger than the Moon: two of Jupiter's family of twelve are about 3,000 miles across, as well as one each in the retinues of Saturn and Neptune. But the Moon is exceptional in one way. Although not particularly large on the satellite scale, it is certainly unusually significant compared with its primary. Its diameter of 2,163 miles is over a quarter of that of the Earth (7,927 miles), and the two form what many astronomers look upon as a double-planet system. This immediately prompts the question: Did the Moon ever form part of the Earth?

G. H. Darwin, the son of the great naturalist, certainly thought so when he advanced his tidal theory. Darwin supposed that the Earth, soon after its original formation, was in the form of a molten mass spinning much more rapidly than it does today – with a period of between 2 and 4 hours. A thin crust had formed, but it was still sufficiently soft and plastic for this speed of rotation to cause it to bulge alarmingly at the equator, for the same reason that the weights on a governor fly outwards when the motor is started. However, there is a maximum speed at which a solid can rotate and remain whole. A flywheel, if rotated too quickly, literally bursts, and Darwin envisaged the same fate overtaking the primeval Earth. Instead of bursting, however, it developed an unstable projection due to the rhythmic tides imposed on it by the Sun. Once formed it quickly developed, became pear-shaped, and finally fractured. The 'neck' of the pear went spinning away into space as the embryo Moon, which finally formed its own crust, settled into a stable orbit, and cooled into the dead world we see today.

It is a well-known fact that the continental coastlines, if brought together, would fit almost exactly into an extensive land-mass covering about a third of the Earth's surface, and there seems little doubt that very early on in geographical history this primeval continent did indeed exist. What, then, caused it to divide into Asia, America, and Australia? On Darwin's theory the agent was the Moon, whose genesis is now marked by the vast cavity of the Pacific Ocean. In those days the crust was thin and cracked easily, while the plastic layers underneath permitted the fragments to drift apart; subsequently tremendous precipitation from the steamy atmosphere filled the gaps and produced the present-day oceans. We might suppose that had the Moon not been born in this manner, the Earth would be one vast sea and fish would be the rulers of terrestrial life.

Sadly for the romantics, few scientists today are prepared to consider Darwin's original theory. The moon, although large on the planet-satellite scale, has insufficient mass to have broken away from the proto-planet on its own. The idea has been put forward that not only the Moon, but also the planet Mars, was born in this primeval rupture; alternatively, the Moon and the Earth might have been formed separately, but in close proximity, in the initial condensation of the solar system. One thing is certain: the Moon and the Earth were formed at the same epoch. By measuring the amounts of radioactive isotopes in the surface samples brought back by *Apollos 11* and *12*, it has been found that many of the rocks are over 3,500 million years old. The age of one particular sample is put at 4,600 million years, which is very nearly the calculated age of the solar system.

However the Moon was formed, we can at least be certain that in its young days our satellite was much closer than it is now, because of its habit of keeping the same face turned towards the Earth. When it was young it was doubtless spinning in a few hours, but the tides raised on it by the nearby Earth slowed it down and finally trapped the hemisphere we see

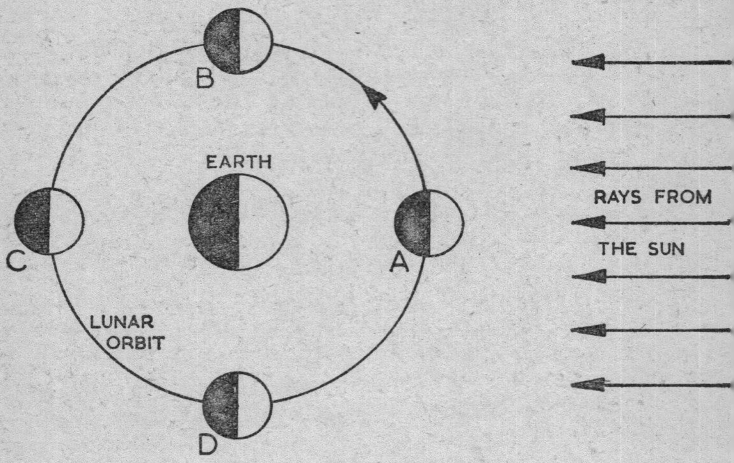

FIG. 4. *The phases of the Moon* (Not to scale).

today in its position of perpetual reverence. It is a fact that this hemisphere has a very slight but positive bulge that could have been produced only when the Moon was still in a plastic form. This bulge is what the Earth's gravitational pull has worked on to slow the axial spin down, until it now revolves once in the $27\frac{1}{3}$ days it takes to go round the Earth. This value is known as the sidereal period.

The phases of the Moon, which arise from this orbital motion, are shown in Fig. 4. When at position A it appears in the same direction as the Sun, and if the line-up is perfect a solar eclipse results; however this does not usually happen, because the lunar orbit is slightly tilted with respect to our own. In the normal position of New Moon it lies either north or south of the Sun in the sky, and since its unilluminated side is turned towards us we cannot see it at all.

As it journeys eastward a sliver of the illuminated hemisphere appears, and a narrow crescent can be seen in the evening sky immediately after sunset (many people refer to this as the New Moon, but they are incorrect). As the days pass the crescent widens until the moon reaches position B. It now forms a right-angle with the Sun and we see exactly half of it lit up; this is known as First Quarter because it has completed one quarter of its total journey. Each quarter occupies about a week.

The terminator (the line separating day from night) now changes its curve and becomes convex as the Moon moves into the 'gibbous' phase. After another 7 days it has reached C; its illuminated hemisphere is fully presented, it appears opposite the Sun and so rises at sunset, and we have a Full Moon. Once again, a perfect line-up will produce a lunar eclipse as it passes through the Earth's shadow.

The second half of the lunation takes it progressively later into the night. The disk becomes gibbous on the opposite side and finally withdraws into Last Quarter, at D, when it rises at about midnight. After this it narrows into a crescent, visible only a short time before dawn, and is finally lost in the morning sky when it has returned to A, giving another New Moon.

The time taken to accomplish a lunation is roughly 29½ days, and the extra two days are due to the slowly-changing angle at which the sunlight shines on the Moon because of the Earth's journey around the Sun. Fig. 5 should make the discrepancy clear. At position A, the Moon is New. 27⅓ days later it has returned to the same position relative to the Earth, but the Sun is not in the same direction; it has to move on to B before another New Moon occurs. This gives us the true lunar month, or synodic period.

On the Earth we see objects mainly either by their colours or by the shadows they cast. There is no colour on the Moon, so the importance of shadow makes the lunar observer very dependent on phase when he wants to observe a particular feature. A crater seen when the Sun is shining low over it has every minute irregularity cast into severe relief, while at local noon it may appear as little more than a light-splotched, featureless blur. For this reason, Full Moon is the worst time to observe if spectacular views are required.

A glance at the Moon with even the naked eye shows that its disk is clearly divided into light and dark areas (thereby providing the face of the Man in the Moon), and early observers interpreted these as continents and seas respectively.

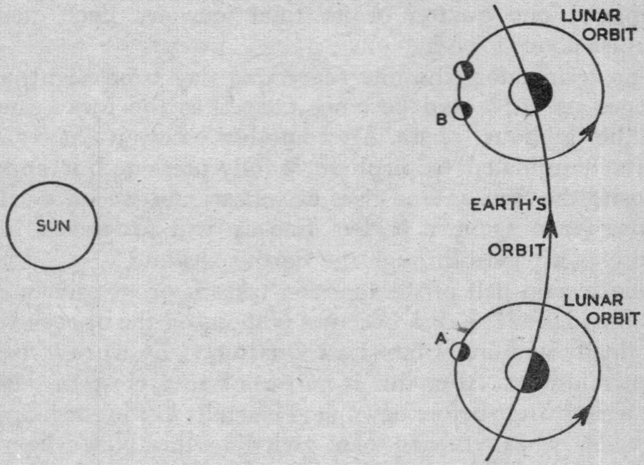

FIG. 5. *The two lunar periods* (Not to scale).

Therefore, although the dark regions are not seas at all, they are still mostly known by the Latin name of *mare*. Scattered over the map (pages 46–47) we find a Sea of Crises (Mare Crisium), Sea of Clouds (Mare Nubium), and even an Ocean of Storms (Oceanus Procellarum). It therefore comes as a shock to discover that these are really dry, dusty lava-plains, almost featureless compared with the crater-encrusted highlands whose stark glory forms an unforgettable sight when seen through a powerful telescope. At the Moon's mean distance of 239,000 miles even a modest instrument is capable of showing craters as small as one mile across, and there are literally thousands much larger than this. Some are so large that entire

English counties could be easily enclosed within their walls.

The lunar characteristic that takes most getting used to is the severe sharpness of everything. On the Earth a mountain range such as the Alps has been eroded by the action of wind and rain; water has frozen and expanded in cracks and split off great masses of rock, and this process is continuing all the time. But on the airless Moon there is no weather to produce erosion, and we see its topography frozen in the forms it assumed at its birth, millions of years ago. Our satellite is a gigantic fossil, virtually insulated from time until the dramatic arrival of *Lunik II* on its surface on September 13th, 1959. By comparison with the lunar mountains the fiercest terrestrial peaks are gentle and accommodating.

This matter of airlessness is so important in lunar affairs that it is worth investigating rather more closely. A moment's thought shows that the atmosphere must be extremely rarefied, because when we view the Moon through a telescope its features are always perfectly clear-cut, without the slightest haziness or blurring. The apparent edge or limb of the disk stands razor-sharp against the sky, and the shadows cast by mountains and craters are perfectly black, whereas an appreciable atmosphere would diffuse light into them.

Other investigations can also be applied. Because the Moon is moving slowly across the sky it must inevitably block out any stars that happen to lie in its path; this phenomenon is known as an occultation. If the Moon possessed a reasonably dense atmosphere the star would dim and redden for some time before its disappearance behind the limb. But this never happens; the star invariably vanishes in the fraction of a second, shining perfectly steadily right up to the moment of disappearance. Solar eclipses would also reveal an air-mantle, and the result of all investigations suggests that if the Moon possesses any atmosphere at all, its density at ground-level cannot exceed a million millionth of that of terrestrial air – which is itself about a million times thinner than the best vacuum we have yet been able to produce in a laboratory!

We must look to the small mass of our satellite for the explanation. It would take 81 Moons to equal the Earth in

Leibnitz Mts.
Mare Australe
Vlacq
Cuvier
Janssen
Maurolycus
Rheita
Furnerius
Piccolomini
Altai Mts.
Sacrobosco
Werner
Azophi
Fracastorius
Petavius
Catharina
Abenezra
MARE NECTARIS
Cyrillus
Abulfeda
Vendelinus
Mädler
Theophilus
Langrenus
Capella
MARE FŒCUNDITATIS
Delambre
W
Mare Smythii
Godin
Taruntius
MARE TRANQUILLITATIS
Agrippa
Ariadæus cleft
Condorcet
Jansen
Plinius
Manilius
MARE CRISIUM
Proclus
Hæmus Mts.
Mare Marginis
Macrobius
Römer
MARE SERENITATIS
Cleomedes
Posidonius
LACUS SOMNIORUM
Caucasus Mts.
Geminus
Gauss
Eudoxus
Atlas
Hercules
Aristoteles
Endymion
Mare Humboldtianum
Thales

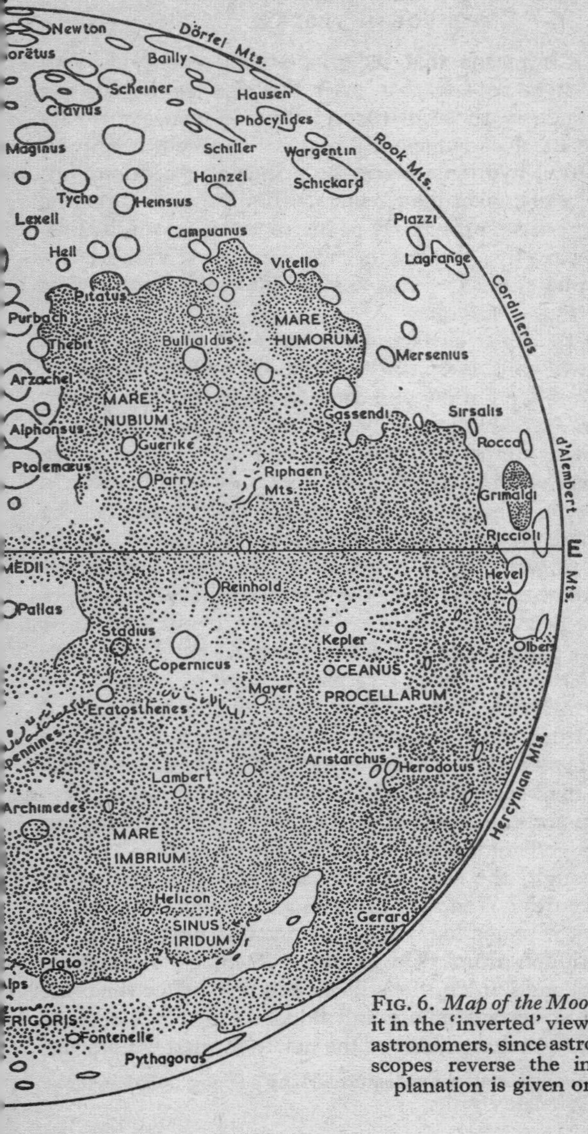

FIG. 6. *Map of the Moon.* This shows it in the 'inverted' view familiar to all astronomers, since astronomical telescopes reverse the image – an explanation is given on page 274.

mass, and this means that its gravitational energy is very much less; it is so low that even the slowest-moving gas molecules, such as those of carbon dioxide, can eventually escape. The Earth is sufficiently massive to retain a grip on most molecules, hydrogen being the principal exception; but when we turn to a giant planet such as Jupiter we find that its atmosphere consists principally of hydrogen compounds. Presumably it was once present in large quantities in the terrestrial atmosphere, but had leaked away into space by the time life had started to emerge.

Although the lunar surface, lying naked to the Sun's blast and the nights' chill, is certainly hostile by earthly standards, space explorers are finding one compensation: the low gravity results in an agreeable loss of weight, as television viewers of the slow, high jumps of Armstrong and Aldrin during the *Apollo 11* visit will remember. This unearthly nimbleness will be found useful when astronauts start exploring the rugged uplands, where craters, valleys, cracks, and pits abound on the small scale as well as the large. Even the seas, though apparently smooth as seen from the Earth with the finest telescopes, have been shown by the *Orbiter* lunar satellites to be covered with depressions just a few yards across. It was once thought that these low-lying lava plains might be covered to a depth of many feet in 'cosmic dust' – the remains of the cloud of particles from which the solar system condensed – but the recent landings have found this layer to be only a few inches thick, with small-scale drifting in some of the craterlets. There may be deep accumulations of dust in parts of the Moon, but the idea of a general blanket has to be abandoned.

Oddly enough, the seas are probably the youngest of all the lunar features.[1] Whatever the manner in which the craters were formed, it seems likely that in its young days the Moon was crater-ridden from pole to pole. Then, as the surface began to cool and harden, the still-molten lava below the crust issued forth at certain points and flooded enormous regions of the globe, re-melting many of the newly-formed craters and

[1] This is actually a matter for dispute. Many selenographers believe the maria to be relatively old.

reducing others to sad shadows of their former state. 'Drowned rings', the rims of circular crater-walls which were submerged almost to the level of the lava sea, are common witnesses of these catastrophic events.

The craters themselves are saucer-shaped depressions, with mountainous walls rising many thousands of feet above the interior and frequently with a mountain mass at the centre of the floor. However, because of the blackness of the shadows when seen with the Sun rising or setting over them, they usually look much deeper than they really are. We may cite the case of a typical crater, Copernicus, in the Oceanus Procellarum, which has a diameter of about 50 miles and walls rising 17,000 feet: truly impressive figures, but anyone standing near the centre of the floor would be unable to see the walls at all. The explanation is simple: the Moon is a small world, and the horizon is only 2 miles away. Copernicus, incidentally, is one of the few large craters to be found on a sea, and is evidently a recent formation.

Craters are named on a system advanced by an Italian priest, Riccioli, who published a map in 1651 on which they were called after scientists or philosophers in general (with a few exceptions, such as Julius Caesar!). This is obviously an excellent system, since there is a convenient and never-ending fund of names. Several well-known amateur astronomers have thereby literally found their way to the Moon's surface, although by this time they have to be content with the smaller and less important formations.

Some craters are truly colossal; the 180-mile Bailly, in the chaotic southern hemisphere, is the largest proper crater, while the nearby Clavius, 150 miles in diameter, contains a string of fair-sized craters running across its floor. In the confusion of certain regions, very few of the formations are perfect; they overlap and distort each other, and some have been so overgrown by later arrivals that they are hardly identifiable at all. Occasionally, on the border of a sea, we find a crater whose seaward wall has been obliterated by lava, while the rest remains and forms a monstrous cliff over the bay.

More interesting, when we come to consider the Moon's

history, are the numerous occasions on which several craters occur close together in a straight line, effectively forming a chain. One of the best known is that on the border of the Mare Nectaris, consisting of Theophilus, Cyrillus, and Catharina, while the huge formations Ptolemaeus, Alphonsus, and Arzachel run down the central meridian. These, and plenty of others, suggest that the agency forming the craters was intimately connected with some local crustal weakness, as against a random process which would be expected to produce craters quite indiscriminately. This, broadly, frames the difference between the meteoric and volcanic theories of crater formation.

Every day millions of tiny solid particles – somewhat larger than cosmic dust – enter the Earth's atmosphere and expire in the streak of fire that many people call a shooting-star; they are more properly known as meteors. Meteors revolve round the Sun like planets, and just occasionally the Earth encounters an unusually large one; the body that fell in Siberia in 1908 must have been several hundred feet across. Once again, these meteors are legacies from the early planet-building processes.

Mercifully large meteors are excessively rare, but when we consider that the Earth and the Moon have been in existence for some 4,000,000,000 years, they must obviously have suffered a great many collisions. A few large meteor craters, notably in the USA and Canada, bear witness to these past impacts, while many must have fallen in the sea. But any landing on the Moon would leave a far more definite mark, since the lack of erosion would preserve the scene of disaster almost as it happened. The meteoric theory therefore suggests that the lunar craters were formed by slow meteoric bombardment during the epoch when the crust was still warm.

The fact that there are many 50-mile craters on the Moon but apparently none on the Earth is overridden by the theory's exponents. For the Earth's atmosphere would very likely cause a large meteor to explode before it touched ground, due to uneven heating and other effects, resulting merely in a shower of smaller meteors; there is also the feeble lunar gravitational pull to be taken into account, for it would allow matter to be flung to considerably greater distances. However, despite atmospheric protection there still seem to be suspiciously few terrestrial craters, and if the meteoric bombardment had been severe enough to pepper the lunar surface we might expect more signs of it among the Earth's geological features.[1] There is also the crater-chain objection, since by no conceivable stretch of the imagination could meteors, separated in time by perhaps millions of years, be expected to produce orderly lines of craters.

The volcanic theory has to cope with the objection that terrestrial volcanoes are nothing like lunar craters, and of course it is obvious that the craters are not in any way comparable to something like Etna. The idea which has received extensive support suggests that when the crust was still very thin, colossal accumulations of gas rose to the surface and raised the crust in the form of huge bubbles which periodically expanded and collapsed until they had built up the circular walls. This suggestion, far-fetched though it may seem at first sight, is supported by the obvious fact that the Moon has had a tempestuous history. Some of its mountain peaks are even higher than Everest, which is no mean feat considering its small size on the planetary scale.

Another piece of evidence may also be cited. Scattered here and there over the lunar surface are to be seen small mounds looking rather like mole-hills – although they are all several miles across. Known as 'domes', their nature has for a long time been a mystery. But the uplift theory accounts for them excellently; they are evidently embryo craters which have somehow solidified in the process of formation.

More mysterious are the winding rills. The most famous, known as Schröter's Valley, occurs near the crater Herodotus in the Oceanus Procellarum, and was discovered as long ago as 1686 by Christiaan Huyghens; at its widest point it is five miles across, and it meanders in a U-shape for over a hundred miles, petering out on the mare surface. Although dozens of these rills were discovered telescopically, literally thousands of narrow ones have been recorded on the *Orbiter* pictures.

[1] We must, however, remember that erosion would have rendered many terrestrial craters virtually unrecognizable.

Formerly thought to be narrow, V-shaped clefts, they are now seen to be shallow and U-shaped, like river valleys. Many of them, too, seem to originate on high ground and 'flow' down the natural gradients, just as water would do if it were released in a sudden flood. Indeed, some investigators believe that regions of the Moon may contain reservoirs of water locked away under a thick layer of permafrost. Although the lunar surface temperature rises to that of boiling water at noon, the surface layer is such a good insulator that practically no diurnal temperature variation at all occurs at depths of just a few feet. Permafrost layers several hundred yards down might survive for aeons until ruptured by meteoric impact or volcanic activity, producing a temporary outflow.

Possibly the most curious of all the lunar features are the rays. These are long streaks of a whitish substance which radiate from many craters, looking rather like the conventional rays drawn around the Sun. Copernicus is the centre of a prominent system, but the most extensive of all belongs to the 54-mile Tycho, fairly near the south pole. Some of its streaks extend for a thousand miles across the surrounding uplands, while others disappear over the limb on to the hidden hemisphere. They are best seen near Full, when they actually obscure much of the detail over which they pass. They are undoubtedly a surface deposit (it looks rather as though some celestial artist had lightly brushed white paint over the rocks), but so far they remain unexplained. The Moon, in spite of its closeness, retains its secrets well.

Crater-building processes ceased probably thousands of millions of years ago. It has popularly been felt that the Moon is a 'dead world'; but the seismometers left on the surface by *Apollos 11* and *12* have recorded a limited amount of internal activity (they were sufficiently sensitive to detect the astronauts' footfalls, as well as minor 'rumbles' that could be due to shallow landslips or meteoric falls). Furthermore, Earth-based observers have from time to time reported apparent activity on the surface, and this question of 'change' has not yet been satisfactorily answered.

Some historical observations of 'change' are no longer

seriously considered. Sir William Herschel, probably the greatest astronomer of all time, announced in 1783 the discovery of three active volcanoes on the dark side of the Moon. When our satellite is visible in the evening or morning sky as a crescent, the unilluminated part of the disk can usually be seen glowing against the sky; this phenomenon, known as Earthshine, is caused by sunlight reflected back to the Moon by terrestrial clouds, and naturally varies according to the meteorological conditions prevailing at the time. It was under these circumstances that Herschel saw his 'eruptions', which were situated near the bright crater Aristarchus, north-east of Copernicus. Aristarchus is often a prominent object under Earthshine conditions, and there is no reason to doubt that Herschel mistook an unusually distinct view of three bright peaks for actual luminosity on the lunar surface.

The case of the small crater Linné, in the Mare Serenitatis, has been more hotly argued. The German astronomer Julius Schmidt announced in 1866 that the appearance of Linné had changed markedly from its former aspect; instead of being a 6-mile crater, it now appeared as nothing more than a tiny pit. Quite recently, some distinguished observers were still arguing about the 'Linné affair'. Now that we have close-up pictures of the area, however, there seems nothing to suggest any recent change. Schmidt, in his original observations, could well have been deceived by unusual lighting conditions. To anyone who has not studied the lunar surface intimately, it may seem astonishing that so drastic an illusion could take place; but we must remember that features are seen by the shadows they cast, and the angle of sunlight is changing all the time. Moreover, each lunation is lit from a slightly different angle, so that it does not necessarily follow that a feature visible at a certain phase will again be visible at the same phase in the following lunation. If the Moon is itself a nearly dead world, the changing illumination gives it a strange life of its own.

Interest in the possibility of lunar activity revived in 1958, when the Soviet astronomer Nikolai Kozyrev, observing with a 50-inch reflecting telescope on the night of November 2/3, noted a reddish glow around the central peak of the crater Alphonsus, near the centre of the lunar disk. He also obtained a spectrogram showing lines due to carbon. His interpretation was as follows: 'First there was an ejection of dust – volcanic ash (appearing reddish in the guiding eyepiece) – and afterwards an efflux of gas (causing the emission spectrum). The effusion of gas could come from magma rising to the lunar surface.'

The effect of this announcement was to stimulate fresh interest in the observation of small-scale, and very temporary, lunar activity, and both amateur and professional astronomers have been conducting programmes to see if transient glows can be detected. Reports are conflicting. Some observers claim to have seen faint reddish glows lasting for just a few minutes, while others, equally experienced, have reported nothing. To many people, the question of these 'transient lunar phenomena' is still very open.

Volcanic stirrings, on a very small scale, may be present, but analysis of the rock samples brought back to Earth has shown no indication at all of any micro-organisms to indicate that the Moon bears life as we know it. This is hardly surprising. Without water, without air, exposed to a range of temperature from about 214° F to −250° F as well as to perpetual bombardment by lethal rays emitted by the Sun – such would be the lot of any form of living matter that tried to exist on the sterile surface. In past ages conditions may well have been different. The Moon must once have possessed an atmosphere, probably consisting mainly of carbon dioxide liberated by the surface-moulding processes, and the moderating influence of this mantle could have made possible the existence of very primitive plant life, perhaps even representing the growth that we now see on Mars. But such advanced creatures as the selenites of H. G. Wells were never very possible (except in the pages of his story, of course), and probably not a living thing has reproduced and died since the Cambrian age of our own planet. The Moon, because of its small mass, ran through its life-history in the time it took the Earth to cool and set the stage for the emergence of life. The last time its now averted hemisphere swung inwards it saw a young, hot, optimistic globe at the end of its baptism of fire.

It is not quite accurate to say that we see only half of the lunar surface. At any single moment we do, naturally enough, but during the sidereal period the Moon both nods in latitude and swings in longitude, so that we see a little way on to the averted hemisphere. This movement is known as libration. The nodding comes about because the Moon's axis is slightly tilted (Fig. 7), while the swinging is the result of the lunar orbit being not quite circular. At its closest (perigee) it is 226,000 miles away, while at apogee the distance is increased to 252,000 miles. (This is a considerable variation, and a perigee

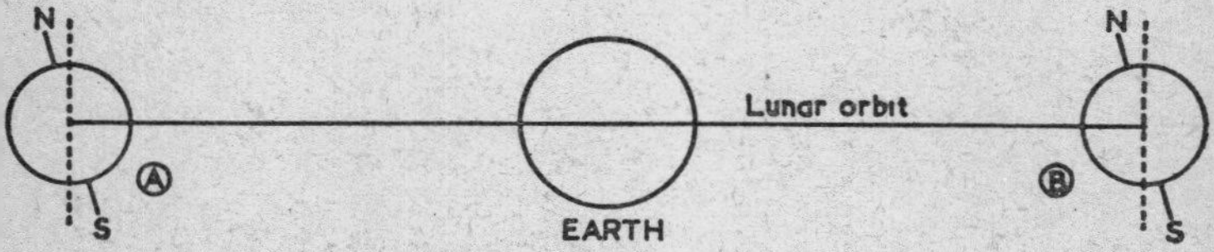

FIG. 7. *Lunar libration* (Not to scale). Since the lunar axis is slightly inclined, and always points in the same direction, it follows that at A we see a little way beyond the south pole, while at the opposite point of its orbit, B, a little of the north is revealed. For purposes of clarity the effect is exaggerated.

Full Moon is appreciably brighter than one occurring near apogee.) In obedience to Kepler's second law of motion (page 64), the Moon must move faster at perigee and slower at apogee if it wishes to remain in its present orbit. But its period of axial rotation, 27⅓ days, remains constant. There is therefore a rhythmic discrepancy between the two motions which causes each limb to advance and recede, and altogether 59 per cent of the surface is presented to our direct gaze. The nature of the Moon's far side had intrigued astronomers for many hundreds of years, until the Soviet reconnaissance probe *Lunik III* sent back television pictures in October 1959. More detailed photographs were obtained by *Ranger 7* in July 1964, to be followed by the magnificent *Orbiter* pictures. We now know that the averted hemisphere is lacking in large seas, and the

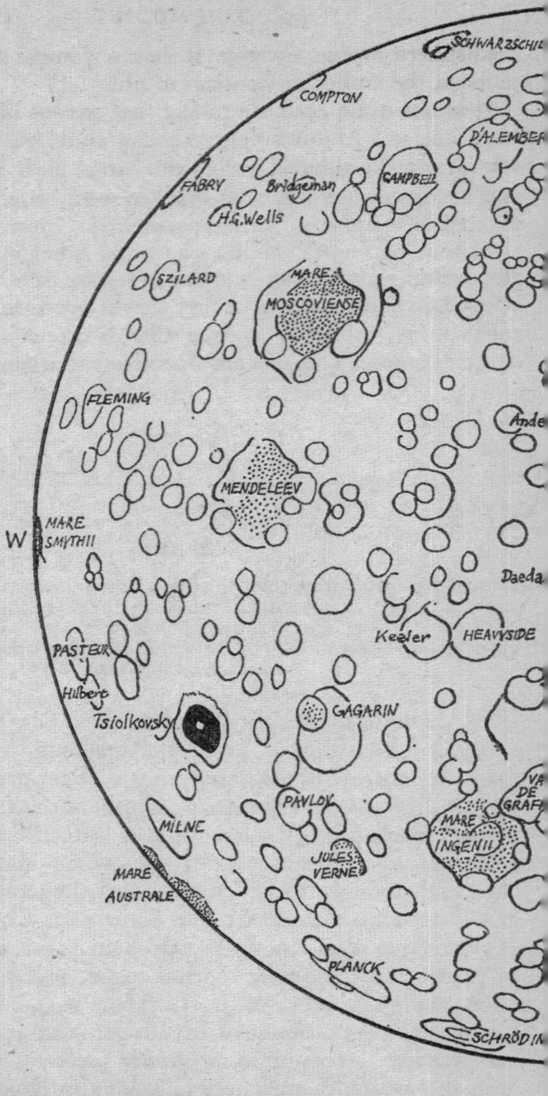
COMPTON
FABRY
Bridgeman
CAMPBELL
H.G.Wells
SZILARD
MARE MOSCOVIENSE
FLEMING
MENDELEEV
W
MARE SMYTHII
PASTEUR
Hilbert
Tsiolkovsky
GAGARIN
Keeler
HEAVYSIDE
PAVLOV
MILNE
JULES VERNE
MARE AUSTRALE
MARE INGENII
PLANCK

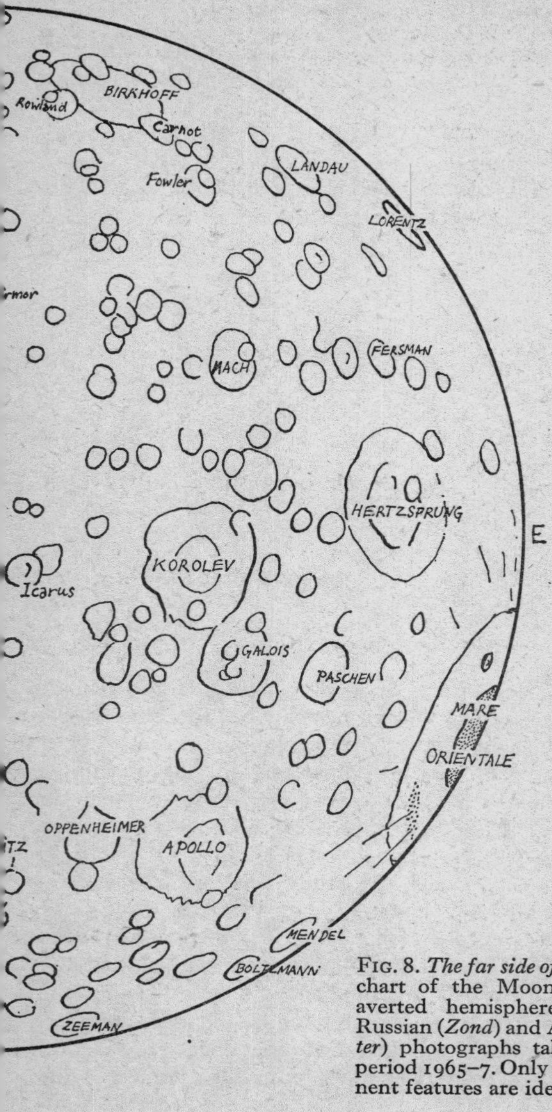

FIG. 8. *The far side of the Moon.* This chart of the Moon's permanently averted hemisphere is based on Russian (*Zond*) and American (*Orbiter*) photographs taken during the period 1965–7. Only the most prominent features are identified here.

singular difference between the two sides indicates that its captured rotation is intimately connected with the moulding of its surface.

Professional astronomers have almost completely ignored the Moon in itself, yet it plays a very big part in the study of solar phenomena by occasionally eclipsing the Sun. This must obviously happen at New, when the line-up is perfect (Fig. 9). Conditions have to be very precise because even when at

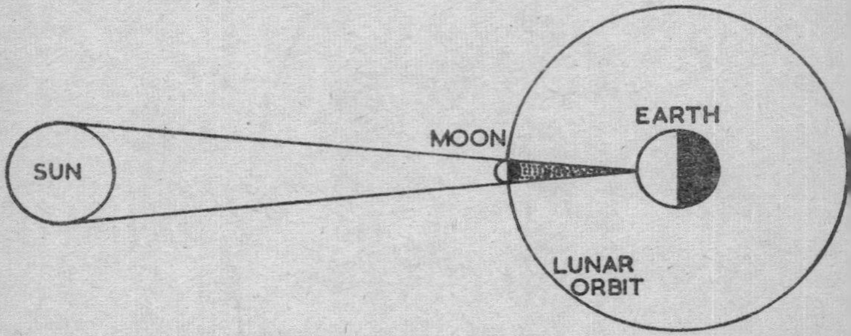

FIG. 9. *Eclipse of the Sun* (Not to scale). Notice that the Moon's shadow only just extends to the Earth's surface. This means that the region of totality of any particular eclipse is very restricted.

perigee the Moon appears only very slightly larger than the Sun, while if an eclipse occurs near apogee the lunar disk is actually too small to cover it completely, and an observer on the central line sees, at mid-eclipse, a thin ring of sunlight surrounding the Moon. Obviously we are very fortunate in this matter, for if the Moon were just a few thousand miles farther away we could never see the corona at all.

Lunar eclipses, in their less gaudy way, are also very interesting. The Moon is eclipsed when it passes through the Earth's shadow, a state of affairs that can obviously occur only at Full (Fig. 10). At the Moon's distance the shadow is some 5,700 miles wide; its orbital speed is nearly 2,300 mph, so that it can remain totally eclipsed for as long as 100 minutes, to say nothing of the time taken to enter and leave the shadow. There is also, surrounding this 'umbra', a much lighter annulus called the penumbra, where the sunlight is only partly cut off. This means that the Moon begins to dim some time before it enters the deep shadow, and the whole eclipse can last for 6 hours.

We might expect our satellite to disappear completely once it has entered the umbra, but in fact it almost invariably turns a deep bronze colour. This is due to sunlight being refracted by the atmosphere and lighting up the shadow; to an observer on

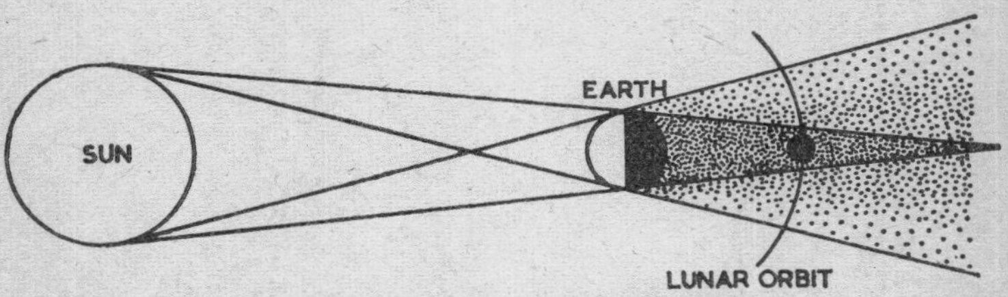

FIG. 10. *Eclipse of the Moon* (Not to scale).

the Moon at the time of a total lunar eclipse, the Earth would be a black disk surrounded by a gleaming reddish halo. It may not be long before human eyes feast upon such a sight.

For about 2,000 years it has been asserted that the Moon and the weather are somehow related; the belief that conditions change markedly at New Moon is a familiar one, and some enthusiasts have gone so far as to relate the lunar phase to the growing of crops. But quite recently, in 1962, came the first very surprising evidence that there does indeed appear to be a connexion between the Moon's phase and world rainfall. Australian and American research workers have discovered, in independent investigations, that heavy rainfall is most likely to occur during the first and third weeks of the cycle, the intervals between First Quarter and Full, and Last Quarter and New, being relatively free from intense showers. Indeed, the statistical evidence can hardly be questioned; the problem is, why should this happen? A theory advanced by a prominent Australian physicist, Dr E. C. Bowen, has received wide support.

Bowen points out that precipitation by a cloud depends on the number of microscopic nuclei it contains; these act as centres for the formation of water droplets, and when they are massive enough they fall as rain. He suggests that these nuclei are not of terrestrial origin, but are simply particles of meteoric dust swept up by the Earth in its motion around the Sun. Evidently the Moon, at certain points in its orbit, shields the

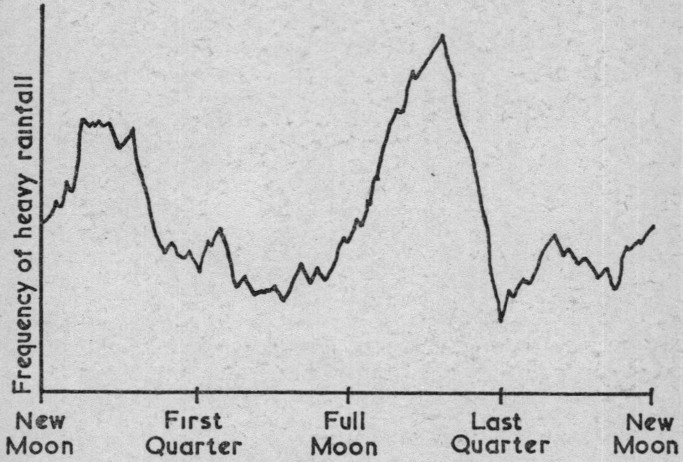

FIG. 11. *The Moon and rainfall.* This is a typical graph among many compiled by American and Australian scientists in their efforts to correlate lunar phase and heavy precipitation, and the curve speaks for itself. It must, however, be emphasized that this refers only to heavy showers and not to total rainfall.

Earth from the maximum intensity of this barrage and therefore reduces the number of rain-forming nuclei.

How are we to explain this blanketing? Simple gravitational attraction would not account for the effect, and we cannot invoke magnetism, since probes such as *Lunik II* have shown the lunar field to be virtually non-existent. On the other hand, it could be due to the phenomenon known as static electricity. If two scraps of paper are charged by running a comb through the hair, they repel each other; and Bowen suggests that both the Moon and the meteoric particles carry an electric charge. In certain positions the Moon therefore diverts the particles clear of the Earth altogether, while in others it is out of the way of the main stream and allows the atmosphere its full complement of nuclei.

There are other explanations also. Many geophysicists believe that the rain-forming nuclei originate on the Earth's surface, being blown into clouds as fine dust. In this case, it is possible that the Moon has some influence on the behaviour of the atmosphere, once again modifying the supply of nuclei. At all events, there can now be little doubt that at least one superstition concerning our satellite contains an element of truth.

CHAPTER 4

The Planets

THE STARS in the night sky, all of which are at terrifying distances from the Earth, stay in virtually the same positions from year to year and even from century to century; if Hipparchus, who mapped the sky 200 years before the birth of Christ, found himself reinstated in the twentieth century, he would have no difficulty in identifying the constellations he observed 2,000 years ago. But in front of this rock-steady pattern ancient astronomers noticed five starlike objects that refused to stay still. Two of them coursed quickly through the sky, always remaining so near the Sun that they could be seen only in twilight conditions, while the other three slowly circled the entire sky. They called them wandering stars, or planets.

We now know the solar system to consist of nine planets. Closest to the Sun is Mercury, whose mean distance of 36,000,000 miles exposes it to the full blast of the Sun's fury. Next is Venus, which under gentler conditions shrouds its true surface beneath impervious cloud. Beyond the Earth is its outer neighbour Mars. Mars, 142,000,000 miles away, is rather like an aged condition of our own planet; but, having lost its internal heat more quickly because of its smaller size, and with only a thin atmosphere to hold in the Sun's warmth, it is an inhospitable world.

These four are often called the terrestrial planets, for they are all rocky globes like the Earth and presumably are experiencing the same basic life-history; given adequate protection, men could probably eke out some sort of existence on any of them. But the case is quite different with the next four, the 'giant planets': Jupiter, Saturn, Uranus, and Neptune. At their great distances from the Sun they are intensely cold, and, what is more, their constitution is violently different. They consist of hydrogen and the poisonous gases ammonia and methane; these swamp their surfaces beneath icy clouds thousands of miles thick, and only the sheerest bravado would incite an astronaut to investigate these virtually unknown worlds.

Moreover, there is another hazard. Jupiter, with a diameter of 88,700 miles, has such a powerful gravitational pull that no rocket likely to be constructed in the foreseeable future could ever escape from its clutches. Saturn, slightly smaller, presents the same problem, and the remoteness of the two outer giants, with diameters of about 30,000 miles, makes the journey unthinkable in terms of present-day speeds. Even light takes nearly 4 hours to cover the 2,790,000,000 miles to Neptune, and a round journey would take literally years.

Marking the perimeter of the solar system is the ninth planet, Pluto, which sets problems of its own. In size at least it is not a giant, and it moves in a very eccentric orbit that at times carries it closer than Neptune. All planetary orbits are in fact ellipses, as Kepler proved in 1609, but for the most part their departure from circularity is slight. Pluto is altogether a very peculiar body.

These major planets are not the only inhabitants of the solar system. In the wide zone between the orbits of Mars and Jupiter circle thousands of very tiny bodies, the minor planets or asteroids. Only a few exceed 100 miles in diameter, and most are far less, so that they are mere lumps of rock by planetary standards. One or two have exceptional orbits that carry them near the Earth, but by and large they are an orderly crowd. As mentioned earlier, they seem to be the gravestones of an abortive planet.

The Earth's position in the planetary sequence divides them into two very unequal groups: those whose orbits lie between the Earth and the Sun (Mercury and Venus), and those which are more remote. They are classed respectively as the inferior and superior planets, and from the point of view of observation the distinction is a very important one. For inferior planets can never stray far from the Sun, while a superior planet is free to roam right across the sky and is therefore much easier to observe.

Three laws describe planetary motion, and they were all discovered by the strange genius Johann Kepler, working with observations made by that most tyrannical of astronomers, Tycho Brahe. Formulated at about the time when Galileo was struggling with his primitive telescope, they are:

1. The orbit of a planet is an ellipse, with the Sun in one focus.
2. The radius vector of a planet (the line joining the planet to the Sun) sweeps out equal areas in equal times.
3. The squares of the periodic times of the planets are proportional to the cubes of their mean distances from the Sun.

Law 1 was the real breakthrough, for it finally split nature from the demands of perfection that had trapped it down the centuries. Everyone has at one time or another drawn an ellipse: a loop of cotton is dropped over two pins stuck through a sheet of paper, and a pencil is drawn round inside the loop. The degree by which the ellipse departs from a circle, known as the eccentricity, depends on the distance between the two pins. These mark the two foci of the ellipse.

The usual ellipse looks something like Fig. 12 (upper), but the orbits of the major planets are not nearly so eccentric; that of Pluto, the least circular, is shown below. By comparison the Earth's orbit is an almost perfect circle. In fact its distance from the Sun varies only from 91,400,000 miles to 94,600,000 miles, and these two extremes of a planet's orbit are called respectively perihelion and aphelion.

Kepler's second law ties these two points in with the planet's speed. The Sun is all the time tending to drag them inwards, in much the same way as a drifting leaf edges fatally towards a whirlpool; and the closer the planet the faster it has to move in order to counteract this force. Therefore near perihelion it is travelling faster than when near aphelion.

The third law (which Kepler jubilantly called his Harmonic Law) may be interpreted sufficiently by saying that the more distant a planet, the longer its sidereal period or year. Pluto, crawling along at only 3 miles per second (as against our $18\frac{1}{2}$), takes 248 Earth years to cover its huge orbit. On the other hand

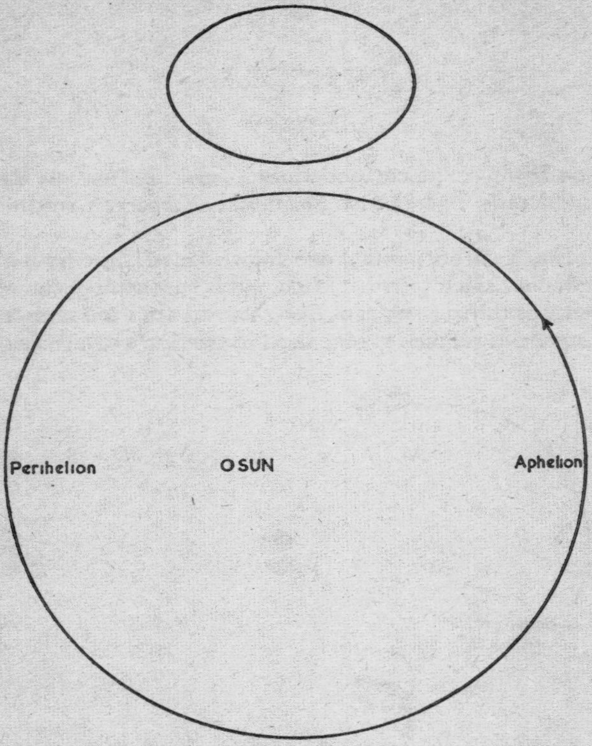

FIG. 12. *Two ellipses.* The upper figure shows a rugger ball; the lower, the most eccentric major planet orbit – that of Pluto. The positions of greatest and least distance from the Sun are called aphelion and perihelion respectively.

flighty Mercury scurries round the Sun in only 88 days at an average speed of 30 miles per second. It is with this tiny bleached world that we must start our tour of the solar system.

CHAPTER 5

Mercury

Mean Distance: 36,000,000 miles *Periodic Time:* 88 days
Axial Rotation: 58·65 days *Equatorial Diameter:* 3,100 miles

IT IS hardly surprising that our knowledge of Mercury is anything but satisfactory; what is more, it is hardly more complete than it was thirty years ago. Aptly named after the messenger of the gods, it is quick to hide itself in the Sun's rays; so eager,

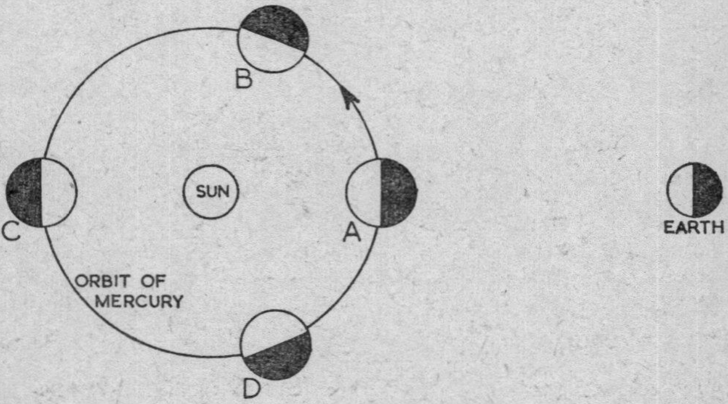

FIG. 13. *The movements of Mercury* (Not to scale). The same state of affairs also applies to Venus. For simplicity, the Earth is taken as being stationary in its orbit.

in fact, that it takes some perseverance to pick it out with the naked eye even a dozen times in a year. In addition, its small disk demands a large telescope to reveal much definite detail. Most large instruments are in the hands of professional astronomers, who are little disposed to chase Mercury when far more profound issues claim their attention.

The movements of Mercury are shown in Fig. 13, and it will be seen that it runs through lunar-type phases. Ignoring, for convenience, the Earth's orbital motion, we begin with Mercury in position A. If the line-up is perfect we see it silhouetted against the Sun, but because its orbit is slightly inclined relative to our own, this does not often happen. Usually it passes north or south of the Sun; its night side is turned towards us, and like the New Moon it is invisible. This position is called inferior conjunction.

A crescent then becomes visible west of the Sun, and by the time it has reached B it appears as a perfect half. This is called elongation, since it is at its greatest angular distance from the Sun (on average, about 23°). It then appears to swing in again, shrinking because of the increasing distance, and becoming gibbous. Finally it reaches 'full' on the far side of the Sun, at C. From this position, superior conjunction, it proceeds to swing out on the Sun's eastern side and run through another apparition in the reverse order of phases.

Although Mercury takes only 88 days to circle the Sun, the Earth's own motion in the same direction effectively slows it down, and the interval between two successive inferior conjunctions is actually about 116 days.

Regular observation of Mercury started with the Italian astronomer Giovanni Schiaparelli, who observed the planet from 1882 until 1889. He combated Mercury's shyness by observing during the day, when both the Sun and the planet were high above the horizon. It is a common misconception that stars and planets can be seen only at night, but this is not strictly true; even a small telescope, if used carefully, will pick out a bright star in broad daylight. It is useless to observe Mercury when it winks near the horizon either before dawn or after sunset, because its low altitude means that we have to view it through the full thickness of the Earth's atmosphere, whose heat currents upset the definition and make a good view more or less impossible. The same is also true of the other inferior planet, Venus.

By this ruse Schiaparelli was not only able to get good views of the planet; he could also observe it continuously for several hours at a stretch. This, as things turned out, had vital consequences, for the faint markings that he managed to glimpse showed no motion at all throughout these periods. The inference was that Mercury's rotation must be very slow. Subsequently he came to the conclusion that relative to the Sun it does not rotate at all; in other words it keeps the same face perpetually inwards, spinning once on its axis during its year of 88 terrestrial days.

There is an obvious precedent here in the Moon's behaviour towards the Earth, but even so Schiaparelli's conclusion came as a distinct surprise to the scientific world. It was, however, quickly confirmed by another planetary observer, Percival Lowell, whose name is mainly connected with Mars. Lowell set to work when the ageing Italian had to retire, tragically, through blindness, and he confirmed the rotation period. Yet there were remarkable differences between the two series of drawings. Lowell, instead of seeing brownish streaks and patches on the planet's rosy disk, drew sharp features looking for all the world like vast cracks or ravines. He saw it as a rocky dead world, seamed and split by the Sun's scorching rays.

It is certainly true that the surface of Mercury, exposed to regular heating and cooling on a fearsome scale, would tend to break up. Yet doubts were immediately cast on Lowell's drawings, and the vast bulk of subsequent work has dismissed his 'cracks' as non-existent. One of the main contributors was the Greek astronomer E. M. Antoniadi, who used the 33-inch refractor at Meudon Observatory between 1920 and 1940 and not only confirmed Schiaparelli's features, but also added some of his own. Two other leading planetary observers, Audouin Dollfus and Henri Camichel, using the 24-inch refractor of the Pic du Midi Observatory in the Pyrenees, provided supporting evidence for the generally-accepted Mercurian features.

All seemed well until the publication in 1965 of radar observations made by G. H. Pettengill and R. B. Dyce of the Jet Propulsion Laboratory. By using very sensitive apparatus, it was possible to reflect a radar transmission off the surface of Mercury and receive the return signal; by measuring the change of frequency of the reflected beam, the speed of rotation could be deduced, and a knowledge of the planet's diameter immediately led to a value for the rotation period.

Pettengill and Dyce announced a sidereal period of 59 ± 3 days. Subsequent analysis of drawings and photographs dating back over the last 90 years, as well as dynamical studies of the system, now strongly suggest that the period is 58·646 days, or exactly two-thirds of its orbital period of 87·969 days.

An indication as to the nature of Mercury's surface came through the work of the distinguished French astronomer Bernard Lyot, who died in 1952. Lyot's approach was to measure the way the total light reflected by the planet varied

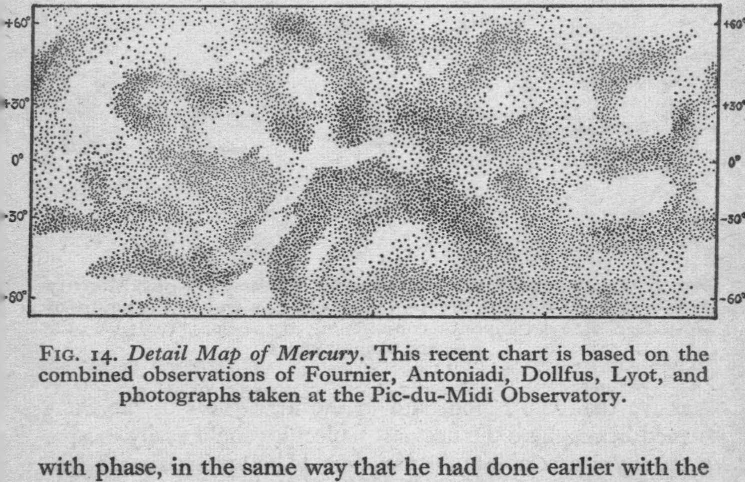

FIG. 14. *Detail Map of Mercury.* This recent chart is based on the combined observations of Fournier, Antoniadi, Dollfus, Lyot, and photographs taken at the Pic-du-Midi Observatory.

with phase, in the same way that he had done earlier with the Moon. It is obvious that when Mercury is at superior conjunction the sunlight is reflected squarely back to us, while at elongation the bulk of the light suffers a 90° reflection. As it moves into the crescent phase the angle becomes still greater (Fig. 15).

If Mercury had a very smooth surface, the angle at which the light was reflected would have very little effect on the total brightness, since the surface would scatter light impartially in all directions. But a rough surface would result in a falling-off in transmission with increasing angle, since the irregularities would cast shadows and so prevent the regions near them from reflecting any light at all. This behaviour, Lyot found, was almost exactly the same as the Moon's. It therefore follows that Mercury is a mountainous world, and the dark patches may well be relatively featureless plains corresponding to the lunar seas.

It has for a long time been a point of dispute whether or not Mercury possesses an atmosphere. It is obvious that if there is any air at all it must be very thin; the planet is hardly more

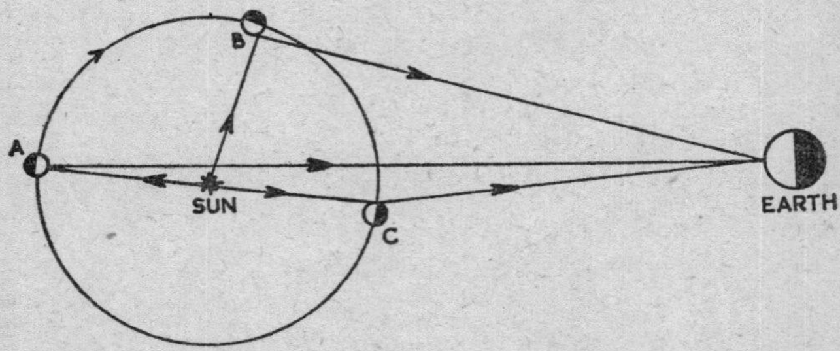

FIG. 15. *Lyot's investigation* (Not to scale). The way in which Mercury reflects sunlight through different angles gives a clue to the nature of its surface. A: near superior conjunction; B: at elongation; C: near inferior conjunction.

massive than the Moon, and in the fierce heat of the Sun-turned hemisphere the heaviest molecules could easily escape. Also, any appreciable atmosphere would betray itself as a bright ring of light around the black body of the planet when it transits the Sun. Transits, which must obviously occur at inferior conjunction, are rare – the last was on May 9th, 1970, and the next will be on November 9th, 1973 – but they always show the planet as a hard and clear-cut round spot. Things are very different in the case of Venus, which shows a clear atmospheric aureole.

Antoniadi, however, believed in a thin atmosphere, and this was to some extent confirmed by the work of Audouin Dollfus. Dollfus, working under the very favourable climatic conditions of the Pic du Midi Observatory in the Pyrenees, which is to date the highest observatory in the world, examined the spectrum of Mercury by comparing it with that of the Sun. We see Mercury by reflected light, and if it has an atmosphere the gases in it should leave their mark on what otherwise is simply the solar spectrum.[1] Dollfus thought he had discovered evidence of a very thin atmosphere, with a density perhaps 1/1,000th of our own. However, this delicate work is made extremely difficult by Mercury's refusal to show itself in a truly dark sky. Normally there is no way out of the difficulty, but during the favourable solar eclipse of February 15th, 1961, the Soviet astronomer Nikolai Kozyrev devoted himself to photographing the spectrum of Mercury during the precious moments of totality, when the sky was relatively dark. This observation shows no trace of an atmosphere, and leads to the logical conclusion that the innermost planet is a virtually airless world. On the other hand, Kozyrev himself has very recently (July 1963) announced conflicting results, resulting from continued spectroscopic work, in which he states that Mercury may have an atmosphere of hydrogen with a ground pressure of anything up to 7 mm., or 1/100th of the terrestrial density. He proposes that it is not a genuine atmosphere, but consists of gas expelled from the neighbourhood of the Sun trapped by the planet's gravitational field; it gradually leaks away, but there are always fresh supplies to replenish the mantle. However, this observation remains unconfirmed, and the question of the Mercurian atmosphere is still a matter of great uncertainty. Some astronomers feel that the only possible constituent of a Mercurian atmosphere is the heavy inert gas argon, which, unfortunately, has no lines in the visible part of the spectrum and so cannot, at the present time, be confirmed.

The Mercurian sky would be curiously alien to a visitor from the Earth. Because it spins so slowly the constellations remain above the horizon for over a month at a time, and against this steady tapestry move two brilliant planets: Venus, shining with

[1] This observation is made even more difficult by the Earth's atmosphere, which imprints its own lines on the spectrum of Mercury.

unprecedented brightness, and a bluish star with a fainter white neighbour – the Earth itself, accompanied by its faithful Moon. But Mars and the more distant planets shrink into the outer darkness, and the sight of his home planet will only encourage our traveller to leave this dead and sinister world unexplored.

CHAPTER 6

Venus

Mean Distance: 67,200,000 miles *Periodic Time:* 224 days
Axial Rotation: 243 days *Equatorial Diameter:* 7,700 miles

VENUS IS the most tantalizing of all the planets. Observationally the same problems apply as with Mercury, except that its elongations are much wider (it can appear up to 47° away from the Sun), and it is therefore sometimes visible in a truly dark sky. At such times it blazes like a miniature lamp, and under good conditions it can cast a perceptible shadow.

This dazzling brightness is due to its cloud-laden atmosphere, which has so far defied all attempts at penetration. Telescopically it appears like a gleaming Moon; dusky features are often visible, but they are obviously nothing more than cloud formations and give no clue at all to surface conditions. For the time being we have to admit that despite its closeness (it can approach the Earth to within 25,000,000 miles) we know relatively little about this neighbour world.

Our knowledge of the suface conditions of Venus is based largely on the results of space probes. The first to reach the planet's vicinity was *Mariner II* in 1962. This was followed in October 1967 by the Soviet probe *Venus 4*, which reached the surface, and *Mariner V*, which passed near the planet a few days later. Two more Soviet probes, *Venus 5* and *6*, hit the planet in May 1969. However, these reconnaissances, valuable as they are, are necessarily confined to a period of just a few hours every 19 months (the period between successive launch 'windows', when a journey to the planet can be accomplished with minimum energy and the greatest likely accuracy). Let us therefore review the work of Earth-based observers first. Considering the difficulties, they have been extremely patient with the cloud-swathed face of this uncharted world.

Venus, in fact, was in at the very dawn of telescopic observation. Shortly after finding Jupiter's four great satellites, themselves elegant evidence for the Copernican Sun-centred system, Galileo turned his telescope to Venus and to his great amazement saw it slowly passing through the lunar phases; from appearing small and nearly round, the phase lessened to half and finally crescent while the disk expanded as it approached the Earth. Finally it swung through inferior conjunction, to reappear on the other side of the Sun and run through the phases in the reverse order.

The bewildered scientist was not slow to realize that here was explosive material indeed. The Ptolemaic or Earth-centred system, which was then vying with Copernican principles, demanded that Venus should always appear as a thin crescent, since it remained in the region between the Earth and the Sun. His observations were therefore confirmation, if any were needed, of Copernican theory. Of course, for Galileo to have blatantly announced his findings would have been the sheerest heresy; he was already in trouble with the Church over his radical views, and the culmination came in 1633 with his appearance before the Inquisition. But Venus herself patiently went on performing her revolutions, and as more and more telescopes were pointed to the evidence so the great Ptolemaic landslide began. The Planet of Love had certainly implanted anything but affection between astronomers and the Church.

Visual observation down the $3\frac{1}{2}$ centuries since Galileo's time has undergone surprisingly little revolution, for telescopic power counts for less with Venus than is the case with the other planets. The reason lies in the very vagueness of the markings. Even when well seen the dusky regions are very difficult to define accurately, and they are always on a planet-wide scale; there is no question of detecting fine detail as on the Martian and Jovian disks. We might sum the position up by saying that while a large telescope will give a more definite view, it will not necessarily show any more than a smaller one: a fact of great comfort to the amateur.

Given that the dusky patches shift and change and are clearly atmospheric features, the main problem visual observers have had to tackle is the length of the day. With Mars and Jupiter there is no problem at all; their disks are a mass of detail, and it is an easy matter to time the return of a certain feature to the central meridian. But the hazy, nebulous disk of Venus offers nothing of assistance. Estimates, if we can call them that, range from 22½ hours to 225 days (which means that, like Mercury, it must keep the same face towards the Sun), and this rather large disparity shows exactly the difficulties that we are up against. This line of attack is clearly hopeless, and resource must be made to less direct methods.

The spectroscope offers one outlet by making use of a phenomenon known as the Doppler Effect. If a source of light is approaching the observer the velocity of the source increases the number of waves reaching the observer in any given interval of time, and the result of this is to shorten the wavelength of the light as he sees it. Therefore the lines in the spectrum shift by an amount corresponding to this shortening, which means that they are all slightly nearer the blue end of the spectrum than they would be were the source stationary. This is called a blue-shift. Should the source be receding, the result is a movement of the lines towards the opposite end: a red-shift. Later in this book we shall see how the red-shift is of enormous consequence, not just in the solar system but in the universe as a whole.

The Effect is easily observable in the case of the Sun. It is spinning from left to right, as we see it, so that if the spectroscope is directed towards its east (left) limb there is a marked blue shift; conversely, the receding western limb gives a red shift. By the same reasoning, the opposite limbs of Venus should also reveal a shift if the rotation is fast enough – about 2 days or less. But so far they have refused to reveal any shift at all, and this therefore rules out the idea of a normal period. It is hard to account for this. All the major planets, except Mercury and apparently Pluto (which is a very odd world altogether), have rotation periods of less than 25 hours, so why should Venus be exceptional in this respect?

If we can dismiss the short period, we can also be fairly sure that the 225-day period is inadmissible. If Venus really kept the same face towards the Sun, its night side would be bitterly cold – even though atmospheric conduction of heat from the day side would raise it above the temperature of Mercury's dark hemisphere. But measurements of the day and night temperature show a relatively small temperature difference, which means that all parts of the globe must periodically face the Sun. What value can we choose in between these two extremes?

Until recently the usual compromise was 'a few weeks', but two lines of research have lately influenced thought. The first is the failure of the space probes to detect any significant

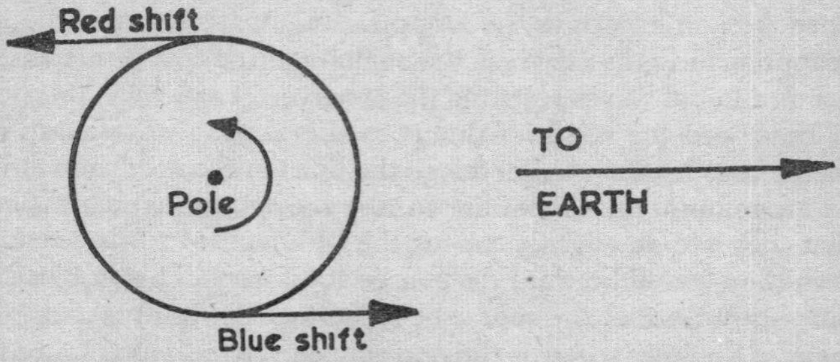

FIG. 16. *Venus and the Doppler Effect.* If we had an equatorial view of Venus and it were spinning rapidly (in just a few days), the approaching and receding limbs would show blue and red spectral shifts respectively.

magnetic field; and a planet's magnetism seems to be closely related to its spin. We may sum it up by saying 'The shorter the day, the stronger the field'; and, if Venus really has no magnetic field, this is good evidence that its rotation is very slow.

The radar method, as applied to Mercury, is more promising. Observations of the planet when near inferior conjunction started in 1962 at the Jet Propulsion Laboratory's station at Goldstone, Texas, and by 1967, when four consecutive conjunctions had been observed, a rotation period of 243 days was announced, with a probable error of only 4 hours. A more recent reduction of the same data has indicated a period of 242·982 days, with an error of ±0·04 day. During this work, a number of discrete regions of very good reflectivity were detected on the planet's disk, and these could be used as datum points for rotational measurements. These reflective regions are known to be rough areas on the surface of Venus, but whether they are truly mountainous, or are simply boulder-strewn fields, is not yet known.

This sensational breakthrough, achieving in a very few years what optical observers have tried to do for 3½ centuries, has caused some not-surprising controversy. But Earth-based studies of the cloud markings, which are all that can be seen by the visual or photographic worker, certainly suggest that the planet's atmosphere is subject to considerable turbulence. Photographs taken in ultra-violet light, which records the tops of the clouds, often show streaky dark markings lying in parallels of latitude, which would be expected if the planet's spin were fairly rapid. Recent work at the Pic du Midi and New Mexico State Observatories has shown that cloud patterns photographed in ultra-violet light have a tendency to recur in a period of between 3 and 5 days, while the observed drifts during a single observing session can imply wind speeds relative to the surface of up to 200 mph! This order of velocity has also been suggested by recent spectroscopic work, using improved techniques to detect the rotational Doppler shift. So we have a puzzle on our hands, and it lies with future observers to explain this large discrepancy between the rotational velocity of surface and upper atmosphere.

Bound up with these matters is the problem of the axial tilt. Most of the planets spin more or less erect – the Earth's axis is tilted from the vertical at an angle of 23½° – but Uranus is an exception, with an axis lying almost in the plane of its orbit.[1] This curious behaviour has prompted the suggestion that Venus also has an abnormally large axial tilt. This is not pure surmise; Dr Gerald Kuiper, one of the world's leading planetary authorities, believes the axis to be tilted at an angle of 85°, and some other observers have come to the same conclusion. This could have an important bearing on Doppler shifts and other measurements, so that in a sense we are almost

[1] The term 'vertical' takes as its horizontal reference the plane of the Earth's orbit. There is, of course, no absolute 'up' or 'down' in space.

back where we started. The cloudy atmosphere defies us implacably.

A recent investigation into the axial tilt was made not by interplanetary probe, but from the relative homeliness of the 200-inch observatory. During the planet's close approach in the autumn of 1962 three Caltech research workers, B. Murray, R. Wildey, and J. Westphal, 'scanned' different parts of the planet in infra-red light, taking advantage of a small window in the Earth's atmosphere that lets through rays at about 10,000 Å. These are invisible to the eye, of course, but they leave their mark on a photographic plate. Since these rays are closely associated with heat-waves, the results are clearly of great interest; they indicate that the planet's axis lies in the normal vertical plane, and that its equator is some 20° F warmer than its poles, though they could not reach any conclusions as to the actual temperature they were measuring. Moreover, this difference refers to the upper atmosphere, and not to the layers close to the surface. However, JPL work confirms that the axis of rotation is a normal one.

Our knowledge of the atmosphere of Venus is dependent very largely on the Soviet probes *Venus 4*, *5*, and *6*, all of which actually entered the planet's atmosphere. Although the readings they sent back to Earth do not entirely agree with each other, we can be sure of three things: the atmosphere consists almost entirely of carbon dioxide; the ground temperature is very high, probably over 300° C; and the atmospheric pressure is at least 25 times that at the Earth's surface. The latter factor was demonstrated forcibly enough when each probe in turn collapsed during its parachute descent through the atmosphere, even though their casings were designed to withstand pressures of up to this amount. Since the altitude of the probes, at the time of collapse, was between 7 and 16 miles, the ground pressure has been estimated as anything between 60 and 140 atmospheres!

Earth-based observers have known for many years that the atmosphere of Venus contains carbon dioxide, the dark bands of which are a prominent feature of its spectrum. It was hoped, however, that free oxygen might be present in the lower, invisible layers; while research also suggested that the upper clouds might consist of ice crystals, analogous to terrestrial cirrus. The temperature of a planet's atmosphere decreases with height, and the upper layers of the mantle of Venus are as cold as —40° C. Unfortunately, the Soviet probes detected no positive trace either of oxygen or water vapour, their analysis being between 93% and 97% carbon dioxide, and between 2% and 5% nitrogen and inert gases, with oxygen accounting for less than 0·4%. It now seems clear that Venus, in its own way, is as hostile to human life as are Mercury and the giant planets, and men will be in no hurry to go there. In fact, the results of these investigations are so discouraging that it may be some time before further space probes are launched in the direction of our neighbour world; Mars beckons more openly, and it is here, rather than on Venus, that men will begin their planetary exploration.

What of the surface itself? Speculation that it might be a planet-wide desert, or a steamy ocean, must now contend with evidence from the Soviet probes that elevations of at least ten miles may be present. There is nothing impossible in this, but this finding must be treated with reserve since it may be due to nothing more than faulty altimeter readings from one of the descending capsules. American radar observations suggest a considerably smoother surface, with no large-scale irregularities exceeding about 1½ miles in height. Certainly, if our interpretation of the cloud markings is correct, and winds greater than anything ever experienced on the Earth are commonplace, erosion in the dense atmosphere will surely have reduced any mountain ranges to the status of low, rounded hills perpetually shrouded by the roaring dust-storms. Such drastic conditions would be fatal to any sort of earthly life. Bacteria, the hardiest living organisms, can survive severe extremes of cold but are quickly killed by excessive heat. It is possible that conditions near the poles may be more bearable, and some physicists have gone so far as to suggest that ice-caps may be present; but in our present state of knowledge such an idea would seem to be optimistic.

Telescopic observation of Venus requires phlegmatic

persistence, but unusual things are sometimes seen. Terminator deformations are one of these, and they usually occur when the planet is in the crescent phase and therefore close to the Earth; they take the form either of a slight projection, or a dent. A projection is caused by an unusually high cloud catching the sunlight when its surroundings are in darkness, while an indentation will result if the cloud is exceptionally low. We see the same sort of thing in the case of the lunar craters, although here the contrast is far more marked. These defects are never very obvious, and require careful attention, but they certainly occur from time to time.

More mysterious are the so-called cusp-caps, which take the form of a brightening at either or both of the cusps. These usually also appear at the crescent stage, and can become very obvious; sometimes they are accompanied by a shady border which isolates them from the rest of the disk. No corresponding features occur elsewhere on the planet, and this very fact suggests that they are something to do with cold currents over the poles – which in turn leads to the conclusion that the planet's axis has a normal tilt. They usually last for several days, or even weeks, and an apparition rarely passes without one becoming fairly prominent.

The strangest feature of all is the visibility of the night side when Venus is a thin crescent. This, the Ashen Light, is only rarely seen; so rarely, in fact, that many astronomers have denied its reality, putting it down to physiological deception. It is certainly alarmingly easy to imagine the whole disk when the planet is in the crescent phase, but too many reliable observers have reported the Ashen Light for it to be an entirely subjective phenomenon. On occasions the dark side really does glow, faint and grey against the sky. It is like a very attenuated earthshine on the Moon.

What can cause this glow? The only sane theory (barring, in other words, a phosphorescent ocean or artificial illuminations) is intense aurorae. Venus is, after all, exposed to much fiercer solar radiation than our own planet, so that the suggestion that its displays of aurorae are more frequent and brilliant than our own is not as far-fetched as it may seem. It has been supported by recent work by Kozyrev, who has detected atmospheric nitrogen in much the same state as that observed during terrestrial aurorae (page 164). He concludes that because of this almost constant illumination the Venusian night sky must be about 70 times as luminous as our own.

Unfortunately, the failure of the space probes to detect a significant magnetic field throws the whole matter open again, for aurorae are closely linked with planetary magnetism; if Venus is really devoid of any significant field, it should also be devoid of aurorae. Yet Kozyrev's observations seem definite enough, and other lines of research have also indicated that Venus is intensely magnetic, with a field perhaps 5 times as powerful as the Earth's! This in turn leads to the conclusion that the planet must have a rapid spin to produce this field. . . . Something is clearly wrong somewhere, but as yet we cannot trace the fault. Indeed, V. A. Firsoff has advanced a theory that the apparent slow rate of rotation detected by radar workers may be caused by a belt of plasma moving in a strong magnetic field, which gives rise to a spurious reflection. He explains the probes' failure to record this field by a continuous stream of electrons emitted by the Sun, passing from the magnetic pole to the equator and cancelling the field to the probes' detectors. At the present time this hypothesis cannot be proved, but it provides food for thought on how very wrong we could still be.

However, all this raises an interesting philosophical point. Suppose that by some miraculous twist we are entirely wrong about the surface conditions; suppose that a race as intelligent as our own lives beneath its swirling, cloud-laden shroud. They could see neither the Sun nor the stars; they could know nothing of the outside universe. Would they assume that it ended in the clouds? Would they build a flying machine to try to discover what was above? In other words, is man's yearning for space a deep-seated instinct, or merely the result of the calculated pulling of the friendly stars?

CHAPTER 7

Mars

Mean Distance: 141,500,000 miles *Periodic Time:* 687 days
Axial Rotation: $24^h\ 37^m\ 22{\cdot}7^s$ *Equatorial Diameter:* 4,200 miles

MARS, PERHAPS because of its warring connotations, has always been the most romantic of the planets. While Venus is the Earth's true twin, as well as its nearest neighbour, Mars is much more of a brother, and an elder brother at that. It is the only planet, with the exception of Mercury, that reveals its true surface to our gaze, and when it swims into the telescope's field its deep ochre disk, jewelled with a gleaming polar cap, is painted with grey-green patches that may possibly indicate the only extra-terrestrial life forms in the solar system.

But Mars, despite its ice-caps and its thin but comforting atmosphere, proves on closer inspection to be a ghost world. Its surface is puckered like the Moon's, the craters partly smoothed out by erosion, and the ashy soil must be so barren that it is becoming increasingly doubtful whether even the lowest known forms of plant life can manage to survive there. If oceans ever existed, they have long ago evaporated away, leaving polar caps only a few inches thick which may, indeed, be composed of frozen carbon dioxide rather than ice. If Mars has not already died of thirst, it is very close to doing so.

The first of the superior planets, Mars presents very different observational problems to Mercury and Venus. Obviously it can never pass through inferior conjunction, and it is actually opposite the Sun in the sky at the time of closest approach (Fig. 17). This position (A) is known as opposition, and its sunlit face is turned directly towards us. Superior planets can never show much of a phase, although at certain points of its orbit Mars appears distinctly gibbous.

After opposition Mars swings slowly away – slowly, because

the two planets are moving in the same direction and the net velocity is reduced – and after some 13 months it has found refuge at the other side of the Sun, at conjunction. Like an inferior planet at superior conjunction it is now unobservable, and in any case its distance has increased so drastically that no

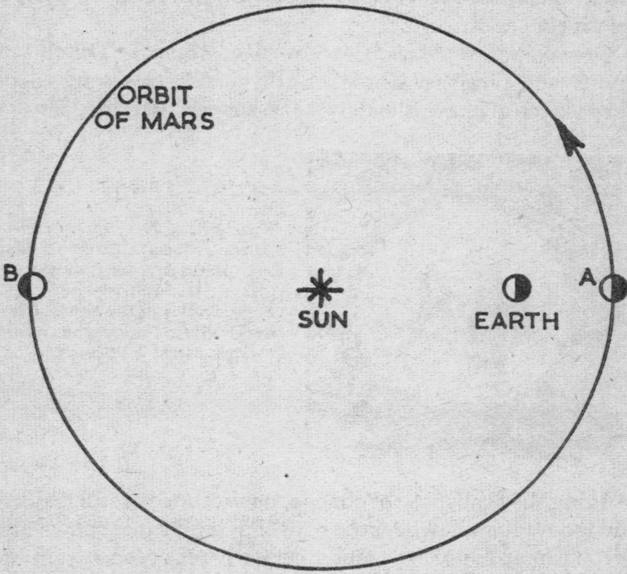

FIG. 17. *The movements of Mars* (Not to scale). For the sake of simplicity, the Earth is taken as being stationary in its orbit. It is clear that Mars is far closer at opposition (A) than at conjunction (B), when it is also very near the Sun in the sky.

useful observations would be possible. It takes another 13 months, giving an average total of 780 days, before it returns to opposition. Venus, by contrast, passes through three elongations in this time, so that Mars presents itself for scrutiny at far more prolonged intervals.

Moreover, there is another point to be considered. The orbit of Mars is appreciably elliptical, and its distance from the Sun varies from 128,500,000 miles to 154,500,000 miles. Consequently, depending on its position, opposition distances can range from 35,000,000 miles to 63,000,000 miles, and this complicates matters still further. A complete cycle of oppositions takes about 16 years; the last perihelic one occurred in 1956, and the next is due in 1971, while at other times we are in the unfavourable aphelic season, when the disk appears relatively small.

Observation of Mars began in 1659, when the Dutch astronomer and physicist Christian Huyghens made a now legendary sketch that clearly shows a V-shaped marking. Since the

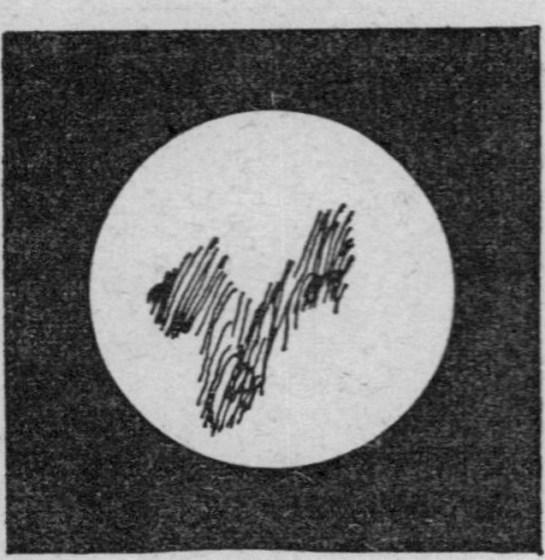

FIG. 18. *The first sketch of Mars.* A facsimile of the drawing made by Christian Huyghens on November 28th, 1659, showing the Syrtis Major and Hellas. Today, the smallest astronomical telescope will give a better view.

markings basically do not change, his feature can still be identified today; it is shown on the map (Fig. 19) and is known as the Syrtis Major. Huyghens and other early observers saw in Mars a true replica of the Earth. What we know now to be deserts they took for fertile uplands, while the darker vegetation areas were interpreted as seas. This idea was slow to die; it was still extant in the nineteenth century, and when Wells wrote his immortal *War of the Worlds* his Martians brought cuttings of their Red Weed with them, which spread in riotous confusion in the more favourable terrestrial conditions.

The present phase of interest started with the industrious Schiaparelli, this time at the very favourable opposition of 1877. Previous observation had been desultory; two observers better known for their lunar work, Beer and Mädler, had published a map in 1840, but since the markings were obviously

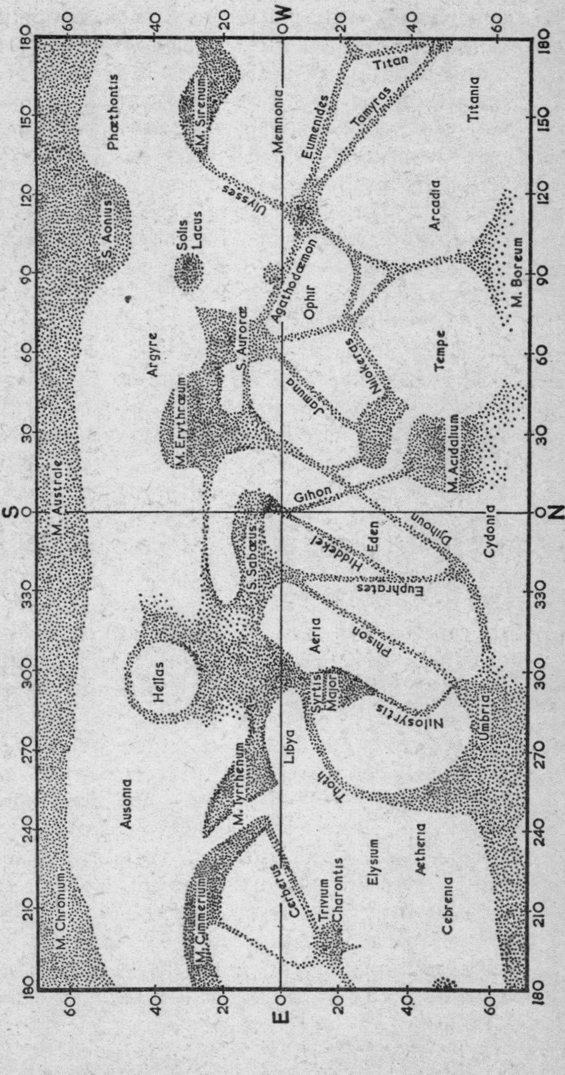

'IG. 19. *Map of Iars*. It is im-ossible to draw definitivechart, ince many fea-ures vary with he seasons, and ome, such as he Solis Lacus, ndergo long-erm changes. Ioreover, dif-erent observers isagree to a sur-rising extent. 'his chart, how-ver, shows an average' repre-entation of the lanet's surface. Jotice how the ero of latitude nd longitude as been chosen t the well-de-ned tip of the inus Sabaeus.

permanent there was a clear need for a chart giving accurate positions for future reference. This is what Schiaparelli set out to do.

But his original intentions were soon distracted by a very remarkable fact. Just occasionally, when the usually turbulent air steadied itself and gave a perfectly sharp image for perhaps half a second, he glimpsed some curious dark streaks running across the disk. It soon became obvious that they were fixed in position; what is more, as he became used to them so they became more numerous. They were nearly straight and very narrow, and looked for all the world like rivers joining the dark areas. He called them *canali*, which in translation means channels, but which soon became known as canals, a name implying artificiality. Undeterred by obvious scepticism, Schiaparelli followed up his discovery during subsequent oppositions. Remarkable things happened. The 'canals' became narrower, straighter, and more numerous still; by 1884 the Martian surface appeared to be seamed with a network of well-organized roads. Something very peculiar seemed to be going on.

The next person to appear on the scene was Lowell, a prematurely retired diplomat who built an observatory at Flagstaff, in the Arizona desert, specifically to tackle the problem of the canals. He worked from 1894 until 1915, and his results not only confirmed Schiaparelli; they went further still. More canals (crossing the dark areas as well); narrower canals; canals that sometimes appeared double; and dark spots which he picturesquely called oases at canal junctions. A glance at one of his maps shows that nature can have had no hand in this conception. Lowell's Mars is as artificial-looking as a map of London – and certainly better organized!

It is worth going into this gradual sophistication of detail rather more deeply, for it underlines a very important principle of observational astronomy. This can best be shown by the following example. Scattered over the sky are thousands of stars accompanied by much fainter companions; they are known as double stars.[1] A case in point is Sirius, the brightest

[1] See page 194.

star in the sky, a faint companion to which was discovered in 1862. This companion had been known to exist, and there had been many fruitless searches in the years beforehand. Yet once it had been identified, the companion was easily seen with the very telescopes that had failed earlier. There are many other instances, and it is of course bound up with the way the eye behaves.

The same principle applies to the apparent progress of detail on Mars. Obviously there had been no radical change during the period 1877–1915. All that had happened was that Schiaparelli and Lowell had become more familiar with their subject. The oases are a case in point. First seen faint and diffuse, they became almost immediately circular black dots. And the canals, initially broad and meandering, took on a spider-web appearance. Lowell himself argued that this refinement indicated their reality; other astronomers turned the argument against him, suggesting that the canals were an *idée fixé*, nothing more.

At all events, Lowell himself felt in no doubt of their objective reality; as the picture sharpened before his eyes he built up a glowing picture of a living planet. Mars is desperately short of water, and most of its surface is desert; its inhabitants therefore had to irrigate vital areas so that crops could be grown, in the same way as the early Egyptians tapped the waters of the Nile. In fact, the mistranslation of Schiaparelli's word 'canali' was correct. What we were seeing was a planet-wide canal system designed to bring water from the melting polar caps to the dusty desert. The Martians were toiling to a man to prolong their doomed existence.

Looking at Lowell's maps, his conclusions are not only sound but inevitable. But unfortunately for his theory his maps are the only evidence for an 'organized Mars' in existence. Even in his day few observers could confirm even his more obvious canals. Antoniadi, using a larger telescope than Lowell, could see only a few vague, streaky markings, and other observers were equally unsuccessful. This was the state of the canal controversy before the American probes, *Mariners IV*, *VI*, and *VII*, took their close-up photographs of the surface.

What is to be made of this mass of conflicting evidence? The main, and regrettable, conclusion must be that Lowell was not a reliable observer (witness his deductions about Mercury); he was certainly no crank, but his eyes must somehow have misled him. It is inconceivable that the features he saw so readily – his total tally of canals was over 700 – should not have been observed by other astronomers using equipment as good as or even better than that at Flagstaff.

But even if we reject the artificial nature of the canals, the lack of agreement over their appearance is still puzzling. Why should some observers see them as broad streaks, others as narrow lines? The reason lies almost entirely in the physiology of the eye, which tends to join up isolated dots and patches into a continuous line. This, at least, is the explanation put forward by Dollfus for the well-established canals. Under excellent conditions at the Pic observatory he has managed to resolve them into a discontinuous pattern, while the fine Lowell-type canals he dismisses as pure illusions. His view has been confirmed in a report issued by Robert Leighton of Caltech, one of the *Mariner* investigators. The report comments: 'Far-encounter photographs have been examined for evidence of Martian canals . . . Although the Mariner pictures are still in a relatively rough form, several previously identified canals appear as well-defined features . . . Other canals appear to be resolved into a sequence of dark patches of varying size and contrast. In some cases the individual dark areas seem unrelated, suggesting that many canals involve the chance alignment of randomly distributed dark patches. Variegated shading has been noted in some of the well-defined canals, but the true physical nature of these features is still unknown.'

Let us leave the canals and return to Mars itself, which, although once thought to be a dried-up version of the Earth, is in our view becoming increasingly lunar-like. It is small, with only a tenth of the mass of the Earth, but this is still sufficient to preserve a detectable atmosphere; and the temperature extremes, although certainly severe, are not impossible. At night, the surface temperature falls to below $-100°$ F, but at midday on the equator it rises to about $+60°$ F. What is more, this temperature can be guaranteed, for the Martian sky is practically free from cloud.

The atmosphere is the crux of the whole matter. It is certainly not breathable; for one thing it is too rarefied. Earth-based observation had suggested a ground pressure of about 1/10 of our own, but the *Mariner* results have shown this to be an over-estimate. The surface pressure is actually about 6½ millibars, or 1/150 of our own, and it corresponds to the terrestrial atmospheric pressure at a height of 20 miles. In other words, visitors to Mars will have to wear pressurised suits just as lunar explorers do, and they will suffer the same encumbrances; for, exposed to this low pressure, the astronauts will find the blood in their bodies to be in a superheated state – and death by blood-boiling cannot be pleasant. Neither will this thin air be of any use for breathing purposes, since its main constituents appear to be carbon monoxide and carbon dioxide; both nitrogen and free oxygen are apparently absent. More disconcerting still is *Mariner VII*'s discovery that ultra-violet rays from the Sun can penetrate in quantities to the Martian surface. On the Earth, these lethal rays are absorbed by a layer of ozone high in the atmosphere, but Mars has no corresponding shroud; and the idea of any earthly type of life existing on its surface is becoming more fanciful as our information increases. Even lichens, the very lowliest type of plant life, will be killed by these rays.

And yet, simple observation suggests that the dark patches (which, incidentally, are warmer than the surrounding lighter regions) might be living. The Martian air is not always clear; just occasionally it turns into a misty yellow haze, undoubtedly caused by winds sweeping the desert sands into colossal clouds which are slow to sink in the planet's feeble gravitational pull. Conditions were exceptionally bad at the favourable opposition of 1956, when for weeks on end the familiar markings were almost lost beneath this temporary veil. Small clouds are frequently seen, and without doubt these dust storms would have long ago obscured the dark regions were they not composed of something living that could push its way to the surface of the new wind-deposited layer.

Furthermore, their apparent link with the Martian seasons forms another piece of evidence. When spring comes to one of the hemispheres and the cap melts, many observers have reported a hardening and spreading of the dark markings. Could it be that these markings are formed of living matter that seizes on the dampness that is carried equatorwards in the leisurely air currents? Perhaps; but opinion is now hardening that the polar caps are composed of frozen carbon dioxide rather than water, and if this is so, and there is no water left anywhere on the planet, our enquiry can reach no rational conclusion.

What of the Martian landscape itself? Earth-based observation had suggested that the surface was relatively smooth, with elevations not exceeding about 3,000 feet, although irregularities in the borders of the polar caps had indicated the presence of valleys or troughs. It was thought, indeed, that Mars might have had a history similar to that of our own planet, once boasting oceans and even fertile land-masses. But the crater-strewn face revealed by the *Mariner* probes makes it clear that Mars has never been a hospitable world. Looking at the remarkable photographs sent back across millions of miles of space, we see a partly-weathered version of the Moon. Of the two hundred pictures sent back to Earth, the vast majority show a landscape similar to the lunar uplands. The dark areas, whatever may be the agent that causes the tint, show craters as prominently as the bright regions, and they extend right up to the pole. In only one area photographed – the region known as Hellas, south of the Syrtis Major – do craters seem to be absent, perhaps being overlaid by tremendous drifts of dust.

As we have already seen, the polar caps themselves are thin – certainly not more than a few feet, and probably not more than a few inches. At maximum extent, they both cover several thousand square miles; when they melt, they seem to leave dry land behind them, and the southern cap, which knows a warmer summer than its northern counterpart, has been known to disappear completely at midsummer. The word 'melt' is perhaps inappropriate, for under the Martian conditions of atmospheric pressure the water or carbon dioxide will sublime directly into vapour, which is why arctic hazes are frequently observed. For the same reason, it seems likely that they are re-formed as snow rather than as ice, although we cannot observe this phase of their history directly, for at this time the polar region is covered with a thick haze of condensing vapour. These hazes are frequently seen elsewhere, especially near the limb, where we view regions under sunrise or sunset conditions. It is not impossible that a thin layer of 'hoar-frost' may occur widely at night, evaporating off in the morning sun and causing the hazes that we see from the Earth.

For most of the time, however, the Martian sky will be a deep blue-black. This is because the thin air is much less efficient than our own at scattering sunlight – on the airless Moon, or out in space, the sky is always black – and at sunset the stars must rush out with a brilliance quite alien to eyes accustomed to the milky terrestrial skies. The Martian heavens will be an awesome sight, and they possess their own morning and evening star – the Earth, with its faithful moon.

The night sky will also contain two other objects of great interest – the satellites. They are both very small, but more than compensate for lack of size by oddity of orbits. The closer of the two, Phobos, is only 10 miles across, and it is a mere 3,700 miles above the surface. This means that its sidereal period is only 7 hours 39 minutes, far shorter than the planet's day! Phobos therefore rises in the *west* and darts across the sky in 4½ hours. What is more, its small disk (⅓ the apparent diameter of the Moon) goes through more than half its cycle of phases in this time.

Phobos is a world under immediate sentence of death. Few planetary or satellite orbits are stable, but in most cases the changes are so extremely slow that they have no practical importance; our Moon, for instance, is slowly spiralling outwards, and thousands of millions of years of this inching will remove it to a distance of 340,000 miles. Finally it will move in again, and may eventually be torn to pieces by the Earth's influence. But the plight of Phobos is more urgent. It has already lived 99 per cent of its life. In about 35,000,000 years it will have moved in so close to Mars that the pull of the Red Planet will

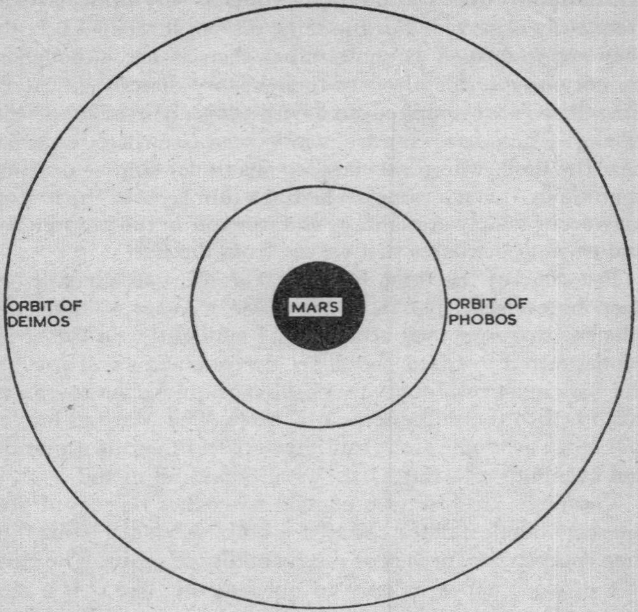

FIG. 20. *The moons of Mars.* From this scale drawing it is clear that both Phobos and Deimos are relatively near their parent. Representing the Earth by the black disk, the Moon would be 15 inches away.

probably shatter it and produce a zone of tiny particles. It is a pity that long before then the last living thing on the planet itself will have succumbed to the insistent rampaging of the desert.

The outer satellite, Deimos, is more sedate than its brother; in fact it is so sedate that it remains above the horizon for 2½ days at a time. It is small and dim, with a diameter of about 5 miles, and at its distance of 12,500 miles its period is 30¼ hours, not much longer than the Martian day. During its period of continuous visibility it accomplishes two complete phase cycles, although its disk is so small that it will require binoculars to show them well. In any case, Phobos at least is markedly irregular (a *Mariner VII* photograph gives it dimensions of 11 × 14 miles), and this will mask the true phase. Also neither satellite will often appear full, since they are so close to Mars that they usually pass through its shadow. What is more, Phobos is so close to the surface that it can never be seen from high latitudes at all.

It is a curious coincidence that Phobos and Deimos should be in such 'practical' orbits; in this happy age of military reconnaissance, few nations would pass up the chance of owning equivalent satellites circling the Earth. Even though it seems a remote chance that a Martian civilization capable of building artificial satellites ever flourished, they should at least prove useful when the time comes to relieve the dying planet of its dreary loneliness.

CHAPTER 8

The Minor Planets

THE STORY of the minor planets or asteroids, the small bodies that circle between the orbits of Mars and Jupiter, began in 1772. J. D. Titius, a professor at Wittenberg, in Saxony, had observed a strange mathematical relationship between the distances of five of the six planets then known. He simply took the numbers 3, 6, 12, etc., and added 4 to each. The resultant series can be matched against the relative planetary distances as follows:

Planet	Relative Distance	Theoretical Distance
Venus	7·2	7
Earth	10·0	10
Mars	15·2	16
?	?	28
Jupiter	52·0	52
Saturn	95·4	100

TABLE I – *Bode's Law*

This discovery was published in an obscure book and failed to receive its due weight of publicity until Johann Bode, a German astronomer, rescued it; it is now rather unfairly known as Bode's Law. Of course this 'law' might be nothing more than a remarkable coincidence, but Bode was sufficiently sure of some underlying meaning to predict a planet occupying the gap at 28. It could only be a small body since it would otherwise be a naked-eye object, so that telescopic searching would be necessary. His suggestion earned additional weight when Uranus was discovered in 1781, for its relative distance of 192 was in excellent agreement with the predicted distance of 196.

The wealthy Baron von Zach, a fellow-countryman, took it on himself to organize a search for this curious planet. It took a long time to get started, but in September 1800 he collected together five other German astronomers at Schröter's observatory[1] at Lilienthal and planned the campaign. Each member of his band of 'celestial police', as he affectionately called them, was to devote his search to a particular small region of the sky, while he cast around for more candidates.

The principle was simple enough. A planet does not stay still relative to the stars because of its orbital motion, and since the undetected planet was faint and therefore must have a small disk, the best chance of finding it was by its motion. Accordingly, the observers carefully mapped the stars in their particular zone at intervals of 3 or 4 days. If one of the 'stars' was in fact the planet, its movement would betray it.

The outcome was ironic. The planet was in fact detected a few months later, on January 1st, 1801; but the discoverer was neither a celestial policeman nor, indeed, an asteroid hunter at all. He was an Italian astronomer, Piazzi, who was patiently compiling a star catalogue, and on the evening in question he came across a 'star' which soon proved to have a motion of its own. It is interesting that Piazzi was down on von Zach's list as a possible co-operator, although he himself was unaware of it at the time! Piazzi followed it for 40 days, when a dangerous illness struck him down and he had to cease telescopic work.

Communications were hazardous even by present-day standards, and by the time Piazzi's letters had reached their destinations the planet had moved into the evening twilight and was unobservable. There was panic in the astronomical world. By the time the Sun had moved out of the region, the newcomer would have strayed too far from its original position to be easily redetected unless an orbit could be computed from Piazzi's observations. It was a task for mathematicians; astronomers, for the time being, were helpless. But nobody seemed equal to the task. September came, by which time the planet should have moved into the morning sky and so be once more accessible, and the celestial police had no clue where to search! Then the problem came to the ears of the young mathematician Gauss, then only 25, who had recently devised a new system of orbit computation. Here was a golden opportunity to try it out. He set to work and by November had published his results.

[1] Johann Schröter (1745–1816) was a famous lunar observer.

Nature now took a hand, and for weeks the skies were covered with impervious cloud and mist. Von Zach's squadron watched anxiously. On December 7th their leader caught a glimpse of what might have been the planet, but the sky clouded over again and it was not until the last day of the year that he managed to glimpse it with certainty. Confirmation followed on the next night, the precise anniversary of Piazzi's discovery. It was a brilliant vindication of Gauss' work, for the planet was almost precisely in the place he had indicated. The recovery of Ceres, as it was named, marked a triumphant mathematical achievement.

The orbit of Ceres indicated a relative distance from the Sun of 27·7, in good agreement with Bode's prediction. This happy state of affairs lasted for precisely 87 days. Then on March 28th the co-recoverer of Ceres, Heinrich Olbers, noticed another 'moving star' in the same region. It was another minor planet, now called Pallas. Now if Ceres had one sister, it could have more; and the celestial police, who had fondly imagined their work complete with the discovery of the anticipated planet, had the first inkling that their task might prove an immense one. Olbers gave the first hint of this when he suggested that Ceres and Pallas might be two fragments of a much larger planet that was somehow disrupted early in the solar system's history.

So the asteroid hunters worked on, and in 1804 and 1807 two more discoveries were made. Juno, the third, was found by Schröter's assistant, and the fourth and brightest, Vesta, was picked up by the industrious Olbers. Vesta can occasionally be seen with the naked eye, at which times it is only slightly fainter than Uranus.

After Vesta there came a long gap, and it is sad to relate that von Zach, Piazzi, and Olbers all died before the surge of discoveries that dates from 1845. In that year, after no less than 15 years of patient searching, an amateur astronomer named Hencke detected the fifth planet, Astraea; and in 1847 he added another one, Hebe. In the same year the disgraceful Continental monopoly was broken by J. R. Hind, who discovered two within two months from an observatory in Regent's Park!

Since then not a year has passed without the list being extended.

The minor planets' biggest windfall came in 1891, when photographic searching was initiated. The stars, which always keep the same relative positions in the sky, appear as sharp points, but minor planets, during the long exposure, appear to trail; instead of leaving a starlike image they reveal themselves as a short line. There is therefore no need to go through the arduous business of comparing separate observations. A glance at the plate will show if there is a minor planet present.

This method occurred to yet another German astronomer, Max Wolf, of Heidelberg, and he discovered his first asteroid (the 323rd) on a photograph taken in December 1891. Thereafter mass discovery began in earnest, Wolf himself finding more than a hundred, and by 1903 the total had passed the 500 mark. Today several thousand have been recorded, many of them turning up on plates exposed for a quite different purpose. Unfortunately it is not enough simply to detect an asteroid; its orbit must be computed so that it can always be found again, and this is an arduous process, so that many of the so-called 'discoveries' are undoubtedly re-discoveries of earlier objects whose orbits were never investigated.

Asteroids are almost invariably given a feminine name.[1] Mythological sources were naturally the first to be tapped, but since the number of well-determined orbits is approaching the 2,000 mark the well ran dry long ago; nevertheless, the christening mania still goes on. No. 387, Aquitania, has obvious connotations, but what are we to make of Photographica, Stereoskopia, Mussorgskia and Pittsburghia? It was rather a shock to find No. 821, Fanny, among this elegant company!

It is the orbits of the asteroids which are of the greater interest, for in themselves they are airless and barren lumps of rock. Ceres, the largest, is only 430 miles across, and a mere handful have diameters greater than 100 miles. Their naked surfaces cannot possibly play host to the lowest forms of

[1] Rather unfairly, exceptionally interesting or important asteroids are always masculine.

terrestrial life (and discussion of any other kind is meaningless), and it is very unlikely, too, that they will ever achieve the space station rôle reserved for them in space fiction.

The vast majority of minor planets circle in the zone between the orbits of Mars and Jupiter, and they keep quite close to the general plane of the solar system; what is more, the average distance is very near the Bode prediction. However,

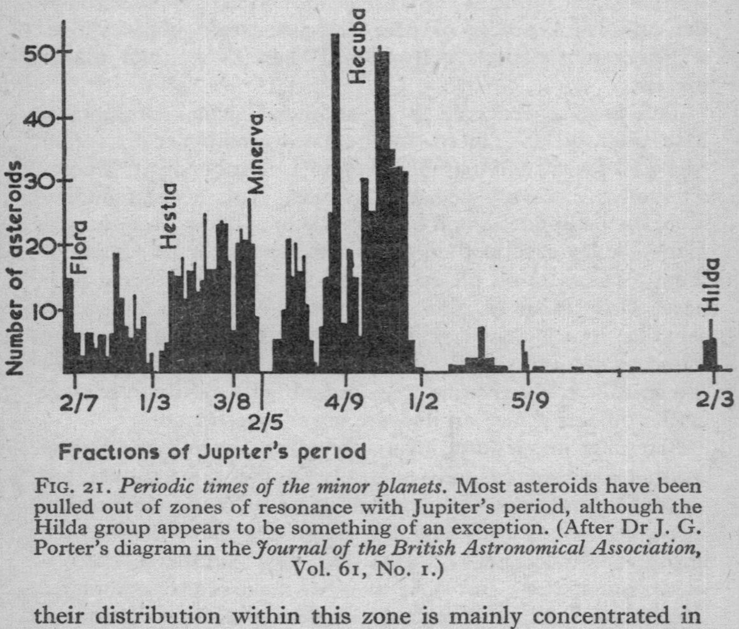

FIG. 21. *Periodic times of the minor planets.* Most asteroids have been pulled out of zones of resonance with Jupiter's period, although the Hilda group appears to be something of an exception. (After Dr J. G. Porter's diagram in the *Journal of the British Astronomical Association*, Vol. 61, No. 1.)

their distribution within this zone is mainly concentrated in various sub-zones, something which comes about through the influence of the giant planet Jupiter.

This is shown very clearly by considering the distribution of the minor planets not in terms of distance from the Sun, but in terms of sidereal period or year, measured against Jupiter's own year ($11\frac{3}{4}$ terrestrial years), and this is illustrated in graphical form in Fig. 21. It will be seen that there are conspicuous gaps at $\frac{1}{2}$, $\frac{2}{5}$, and $\frac{1}{3}$ of Jupiter's year, and at other simple fractions also. Clearly, the giant planet has been responsible for forcing them out of these zones and into others; there is a tremendous group of over 400 with periods slightly less than $\frac{4}{9}$, and there are other smaller families on the brink of

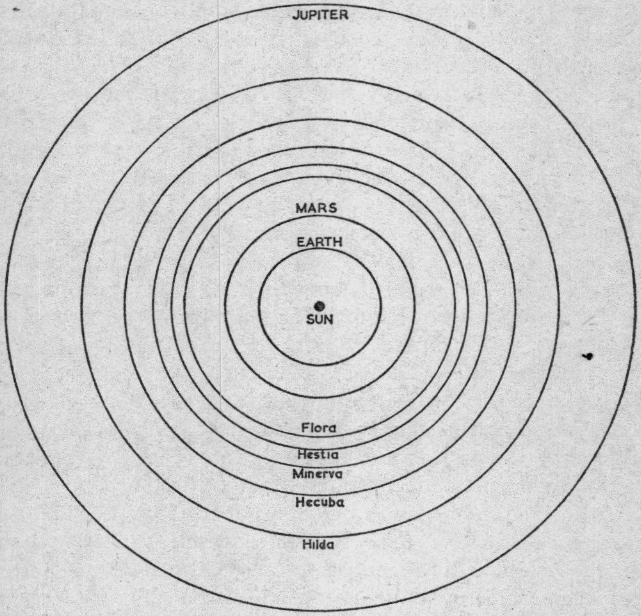

FIG. 22. *The minor planet groups.* Only the most important families are shown here, and the diagram is very schematic, since the individual orbits are so interlocked that were they represented by solid hoops, it would be impossible to lift one without bringing all the others away with it!

the $\frac{1}{3}$, $\frac{2}{5}$, and $\frac{2}{3}$ ratios. They are all called after the main member, and these four are known as the Hecuba, Hestia, Minerva, and Hilda groups respectively. There are many others.

The division into periodic times is of course reflected in their mean distances from the Sun, from Kepler's third law, and Fig. 22 represents the positions of the main groups. However, the diagram is idealized; the orbits are slightly inclined to each other, and they are mostly rather eccentric, so that they seem to intertwine in a bewildering fashion. It takes mathematical treatment to make the groups at all obvious.

Another group is the most interesting of all. Their leader was discovered by Wolf in 1908, and he named it Achilles; it turned out to be unusually large, with a diameter of 150 miles. But its distance from the Sun is the same as Jupiter's! Achilles does in fact move in the same orbit as the giant planet, always remaining some 500,000,000 miles ahead of it; it stays in the same position because both their years are of course the same length. There is therefore no possibility of Jupiter catching it up, and their situation is rather like that of two horses on a merry-go-round. Achilles is a precariously-balanced world.

A little while later another minor planet was detected in Jupiter's orbit; this is Patroclus, which is on the opposite side and therefore follows its master. Subsequently more were added, and the present tally is 12, 6 accompanying Achilles and 4 bringing up the rear with Patroclus. They are known as the Trojans, since their names commemorate heroes of the Trojan–Greek war, but through some mismanagement Achilles and Hector find themselves in the same camp. This is a fine plea for more balanced education in science and the arts.

Not all the minor planets belong to definite groups, and the lone wanderers have their interest; especially those whose orbits are sufficiently eccentric to carry them near the Earth. Eros is one of these. It was discovered in 1898, and has a mean distance of only 138,000,000 miles, which is less than that of Mars. Shortly after discovery it approached to within 30,000,000 miles, but an even more friendly visit came in 1931, when its minimum distance was only 16,000,000 miles.[1] This is considerably closer than Venus, normally the Earth's nearest neighbour, can ever come, which meant that the tiny planet could be used for investigating the Earth's distance from the Sun: that fundamental length of the solar system known as the astronomical unit. All planetary distances have to be measured in

[1] In 1963 the asteroid Betulia came within 15,000,000 miles of the Earth.

terms of this unit, because we have to use the Earth's orbit as a key.

The principle behind the use of Eros is simple enough, though in practice it involved work of the most laborious kind. First of all we find the mean distance of Eros from the Sun, in astronomical units, and to do this we simply measure its

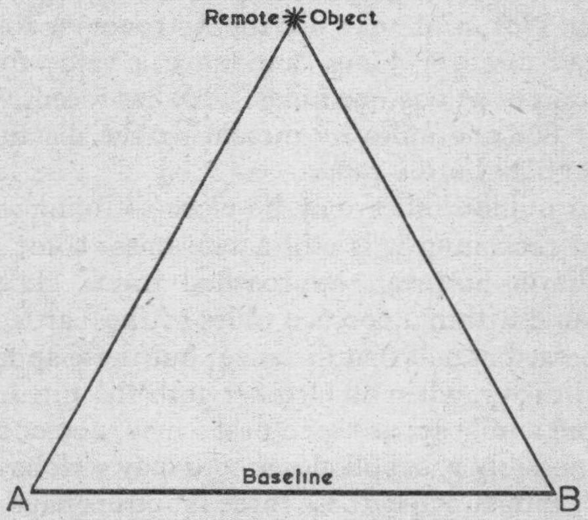

FIG. 23. *The theory of parallax.*

periodic time; the distance follows from Kepler's third law. If we then measure the distance between Eros and the Earth in miles, we obtain a key to the whole unit.

This was done in 1931 by the parallax method, and in theory it is the same as that used by a surveyor who wants to measure the distance of some inaccessible object. A theodolite is pointed at the object and its azimuth, or horizontal bearing, noted. It is then shifted to the left or the right and a new sighting taken (Fig. 23). The difference between the azimuths gives, in terms of the baseline AB, the distance of the object.

The direct application of this method to a planet is complicated by the enormous distances involved; it is hopeless to expect a baseline of a few yards, or even a few miles, to yield much of a shift for Mars or Venus! Also, their disks complicate precise measurement. But Eros was closer than Venus, and showed so tiny a disk that its position could be measured with great accuracy. Accordingly, 24 observatories all over the world joined in the task of photographing the little world during its approach of 1931. Seen against the almost infinitely remote stellar background, its position would shift slightly as seen from different stations, so that by knowing the exact distances between the observatories the distance of Eros itself could be calculated. The actual photographs, however, were only the beginning. Not until 1941 did the Astronomer Royal, the late Sir Harold Spencer Jones, announce a value for the astronomical unit of 93,003,000 miles. This has recently been modified to 92,868,000 miles by measuring the distance of Venus not by parallax but by radar.

Sixteen million miles may be close astronomically, but by terrestrial reckoning it is still a safe miss; some other minor planets have, however, approached much closer. In 1932 Apollo came within 2,000,000 miles of the Earth, and in 1936 Adonis passed at half that distance; but these approaches were eclipsed in 1937, when on October 30th the tiny body Hermes (only about a mile across) sped past a mere 400,000 miles away. It could possibly pass still closer, and may well have done so in the past. Unfortunately these three inquirers have all been lost, because their passages were so fast and furious that normal observing methods were useless. Their recovery will be entirely a matter of chance, and they are all so small that they will be extremely difficult to detect unless they should happen to pass close to the Earth again.

The fact that they can pass through the vicinity of the Earth means that the orbits of these planets must be exceptionally eccentric. Apollo's perihelion distance is less than that of Venus, while Adonis recedes beyond the main minor planet zone at aphelion and swings in almost as close as Mercury. But pride of place must go to Icarus, the Sun-grazer *in excelsis*. At aphelion it is well beyond the orbit of Mars, at a distance of 183,000,000 miles, but its path is so eccentric that perihelion brings it within 19,000,000 miles of the solar surface – far closer than Mercury. Discovered as recently as 1949, Icarus has the most eccentric known orbit; furthermore its plane is tilted at an angle of 23° to that of the planets (Fig. 24).

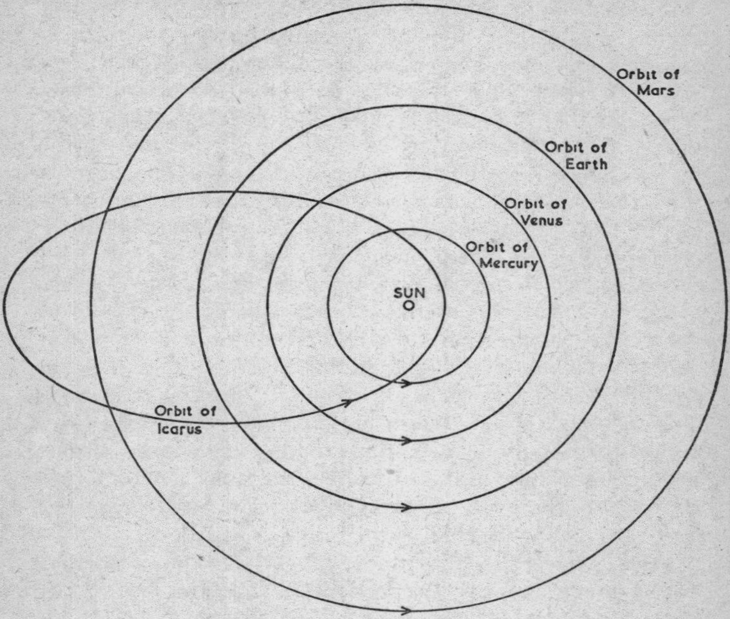

FIG. 24. *The orbit of Icarus.*

Icarus is a tiny body only a mile or two in diameter, but in its orbital caprices it has a larger brother, Hidalgo, which moves in a colossal orbit that carries it from just beyond Mars almost to Saturn. It therefore takes Hidalgo 14 years to accomplish one circuit, while Icarus takes only 400 days. In addition, Hidalgo's path is inclined at the abnormally large angle of 43°, which means that it can stray over a large percentage of the whole sky.

Yet with all these oddities, the vast majority of asteroids have clearly been severely disciplined by Jupiter's attraction. The family groupings form one piece of evidence; another significant feature is that their perihelia and aphelia tend to lie more or less in the same directions as those of the giant planet itself. This means, if we suppose that they were originally distributed in a haphazard fashion, that they must have been formed very early in the solar system's history, for tidal influence is an almost immeasurably slow process.

Evidence is so slight that theories of their formation can be little more than mathematical guesses. Olbers' original suggestion is still held in many circles. Another idea, occurring with the advent of Weizsäcker's planetary theory, is that they are the primordial fragments of a planet that failed to coalesce into a large body, perhaps because of the disturbing effect of the primitive Jupiter. The main query is why the original planet should have been so small – a rough assessment places the total volume of the 70,000 asteroids that are thought to exist at no more than 1 per cent of that of the Earth.

Physically, one or two asteroids have their oddities. Eros has actually been seen to have an irregular shape – it is 14 miles long and only 4 wide – and it spins upon its shorter axis with a period of $5\frac{1}{4}$ hours. Clearly, this produces rapid light-changes with the oscillation of the area presented to the Earth. Several others also show the same light variation, with periods of from 3 to 9 hours, including the brightest, Vesta.

Vesta is something of a mystery. It is the brightest asteroid, but by no means the largest; this means that it must have very high reflectivity – in the region of 60 per cent. Obviously no rock could possibly reflect light so efficiently, and the only reasonable answer is that like Jupiter's satellite Callisto it consists mainly of ice or some other frozen chemical substance. That Vesta should be so unusual in this respect is only one of the problems posed by the minor planets.

CHAPTER 9

Jupiter

Mean Distance: 483,300,000 miles *Periodic Time:* 11¾ years
Axial Rotation (equatorial): $9^h\ 50^m$ *Equatorial Diameter:* 88,700 miles

THE MINOR planets form a marked division between climatic conditions in the solar system. Mars, the outermost terrestrial planet, is cold but not impossibly so; Jupiter, the next major planet, is bitterly chill. The Sun has shrunk to a tiny disk in the black sky, and any astronaut venturing so far from his home planet would find a world swathed in freezing clouds of ammonia and methane. What exist as gases on the Earth's surface are now frozen into liquid or crystalline form.

Jupiter, like its three giant companions, may have a solid surface, but if so it is of academic interest only. The hostile gas layer is certainly thousands of miles thick, and it is so dense that no sunlight could possibly penetrate it. The planet itself is perpetually hidden from our eyes, and all that we can do is observe the disturbances that break out among the cloud features in the upper layers of its atmosphere.

Jupiter is so large that even a small telescope will show a disk (even binoculars distinguish it from a star), despite its normal opposition distance of nearly 400,000,000 miles. Its orbit is not so eccentric as that of Mars, and even when near conjunction it still appears large enough to be observed satisfactorily. This means that it can be followed for a large proportion of the year, so that to the amateur astronomer it is the most satisfying of all the planets.

All the giant planets spin rapidly, and Jupiter has a period of less than 10 hours. Bearing in mind its colossal size – its bulk is twice that of all the other planets put together – this means that a point on the equator is whirling round at 28,000 mph. On the Earth the rate is only 1,100 mph. The resultant centrifugal force has caused the equatorial regions to bulge outwards, and its disk is markedly elliptical; the difference between the equatorial and polar radii is almost 3,000 miles, while on the denser and more sedate Earth it is a mere 13 miles.

Fig. 25 gives a rather schematic representation of Jupiter. The dark belts are clouds, and since they usually hold the same positions they can be given names. In the northern hemisphere the main features are the North Equatorial Belt (NEB); the North Temperate Belt (NTB); and the North North Temperate Belt (NNTB), and the southern hemisphere has its counterparts. But while the belts may be regular in position, they are certainly not regular in appearance. One may fade away completely for several months, while another may divide and appear double. They may also spread in area, encroaching on the neighbouring bright zones. In addition to these general effects, they are always a mass of fine detail which increases in minuteness with improved telescopic power.

Generally speaking the NEB is the most prominent Jovian feature, with the NTB and SEB taking joint second place. However, this is far from always being the case. In 1963 a curious change occurred in the equatorial region, the north border of the SEB and the south border of the NEB extending down to the equator, forming a wide dark zone across the central latitudes. At other times the NEB has faded away and the SEB surged into unexpected prominence; one is never safe in predicting what Jupiter may do in the months to come. The coming and going of the belts, and the constantly-changing wisps of detail, signify disturbances on a truly titanic scale.

Practical observation of Jupiter is mainly a matter of determining the longitudes of the features presented on the disk, and this is done by timing the precise moment at which any particular feature appears to cross the planet's central meridian. The rapid spin produces an obvious shift during an interval of five minutes, and with practice the margin of accuracy can be reduced to half this, equivalent to 1° of longitude. By observing the same feature for several nights in succession, if it survives that long, the error can be reduced to a few seconds.

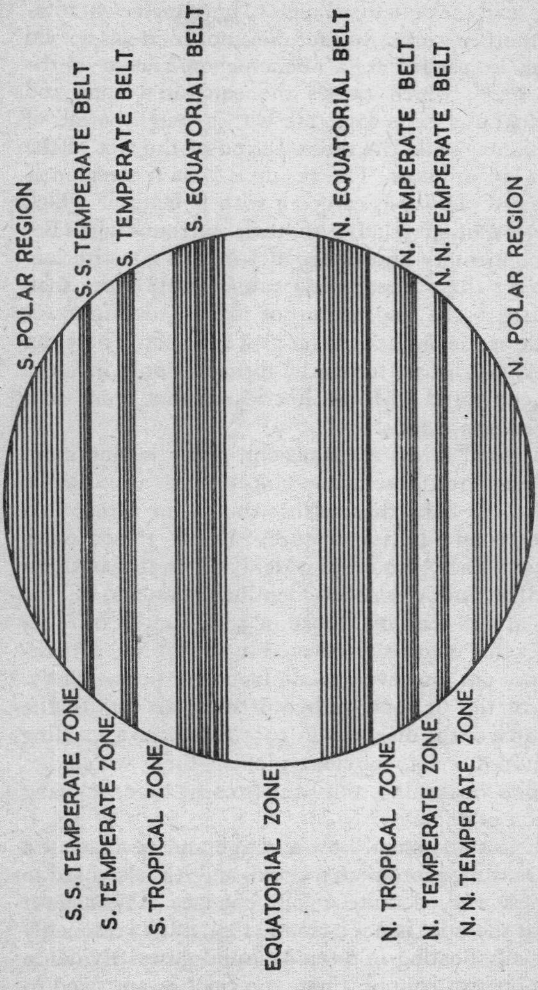

FIG. 25. *Jupiter.* Notice the distinct oblateness of the disk. The belts and zones are, of course, far less cleanly defined than this diagram suggests.

In this way a startling fact comes to light: many features, particularly those in the equatorial region, have their own rotation periods. In other words, Jupiter does not rotate as a solid body. There is, in particular, a phenomenon known as the 'equatorial current', which causes the equatorial zone and parts of the equatorial belts to rotate with a rough period of 9 hours 50 minutes, while the mean period of the rest of the disk is 9 hours 55 minutes. The result is that the region is gradually 'screwed' forward, carrying with it features which themselves have motion relative to their surroundings. Because of our fragmentary knowledge of Jupiter's constitution, theories regarding these phenomena can be little more than intelligent guesses; the suggestion of winds, cyclones, or tornadoes, while vague, is about as far as we can get. There can at least be no doubt that no terrestrial tornado would be more than a breeze compared with the hurricanes that must rend Jupiter's sluggish gas-clouds.

Most Jovian cloud forms are transient, rarely lasting more than a few months, but there is one object which seems to be as permanent as the belts themselves: this is the Great Red Spot. The Spot is an elliptical feature, roughly 30,000 miles long (in longitude) and 7,000 miles wide, lying on the southern border of the SEB, and it has been identifiable, on and off, for just 300 years; it was seen by Hooke in 1664,[1] and it owes its name to the fact that when rediscovered in 1878 it had a brick-red tinge. It does not, however, retain its colour permanently. During much of the present century it merged in with the silvery-cream tone of the disk, but in 1957 there was a startling revival; it acquired a very obvious pinkish tinge which has remained, despite weakening, until the present time, reviving at the opposition of 1962.

It is difficult to understand how so large and permanent a feature can be nothing more than a cloud, and it is also hard to account for these very definite colour changes. The answer seems to be that the Spot is not a cloud at all, but a reasonably solid body actually floating in a sea of liquid gases! Evidence for this comes on two counts. First, the Spot is not fixed in

[1] It is said to be identifiable on drawings made as early as 1631.

position; it drifts from side to side, very slowly, for thousands of miles, and manifestly cannot be attached to the core of the planet. Secondly, its changes of intensity could be accounted for by supposing it to sink a little into the atmosphere (causing a fading and discolouration), and then emerging into relief again.

An interesting sideline is worth mentioning. Suppose the Spot did sink a few hundred miles closer to Jupiter's surface, thereby losing colour. What would happen to its rotation period? The answer is provided by Kepler's invaluable third law: it would lessen, just as a close artificial satellite has a shorter period than a more remote one. Now this phenomenon has actually been observed; at the onset of a period of obscurity, the Spot seems to accelerate its rotation, which is good evidence indeed. It is possible that the same explanation may apply to some other long-lived features as well.

Since Jupiter is representative of the giant planet family, this is a suitable point at which to consider why they are so different from the terrestrial planets. The main reason is their great mass.

The most common element in the universe is hydrogen; it accounts for 90 per cent of the mass of the Sun, and it is only reasonable to suppose that a similar preponderance was present at the genesis of the solar system. Accordingly, when the primeval Earth was formed, it possessed an atmosphere consisting primarily of hydrogen. Hydrogen molecules are fast-moving, and the higher the temperature the faster they move. The Earth, in its early volcanic spasms, was extremely hot. The result was that all the hydrogen leaked away into space, leaving the planet's surface almost naked until the crust cooled and the vast quantities of carbon dioxide and other gases expelled from the interior collected to form the beginnings of its present air-mantle. This explains why oxygen and nitrogen form so considerable a percentage of the atmosphere, while free hydrogen occurs only in traces.

Things were very different when Jupiter came into being. Hydrogen was present just the same, but due to the colossal mass of the planet it was unable to escape. Jupiter therefore managed to retain its primeval atmosphere; the hydrogen combined with other elements to form ammonia (NH_3) and methane (CH_4), while the surplus has been so compressed by the colossal pressure beneath the outer layers that it has ceased to behave like a gas at all; physically it rather resembles a metal! Modern theories suggest that both Jupiter and Saturn have this metallic hydrogen basis instead of the rocky cores possessed by the terrestrial planets.

The densities of the giant planets support this idea. While the Earth has a mean density $5\frac{1}{2}$ times that of water – slightly more than Venus and Mars, and also probably more than Mercury – Jupiter's density is only a quarter as much. Uranus and Neptune are of the same order, while Saturn's density is a meagre $\frac{2}{3}$ that of water. In other words the planet could float in an interplanetary ocean like a colossal beach ball! Clearly, extensive rocky cores are out of the question.

Early observers thought that Jupiter still retained some internal heat, and even went so far as to suggest that it was slightly self-luminous, like a star. However, we now know that the visible 'surface' has a temperature of about $-220°$ F, and the reason for its unexpected brightness lies in the high reflectivity of the cloud-layer. Nevertheless, even such a desperately low temperature is considerably above absolute zero ($-273°$ C or $-459°$ F), which means that the planet is technically 'hot'. Absolute zero is the temperature at which all molecular and atomic motion ceases.

If we take a length of wire and heat it in a flame, it will not start to glow until it reaches a certain temperature; below this temperature it will be emitting heat waves, and we have already seen that radio waves are closely related to heat waves, the difference being that the radio wavelength is longer. It is therefore to be expected that Jupiter should emit radio 'noise', as it is called, due to heating effects, and this emission will give an incidental check on the temperature.

Because of this it was no surprise when in 1956 American radio astronomers picked up this thermal emission from Jupiter, on a wavelength of 3·15 centimetres; it was also received from Venus and Mars, and, later, from Saturn. What was unexpected was the subsequent detection of noise at other wavelengths which could not possibly be related to thermal effects. American and Australian workers have found three main centres of emission at 11, 13·5, and 16 metres, where the intensity suddenly increases and dies away in surges known as bursts. These bursts have posed a most infuriating but fascinating problem.

Leaving aside the question of how the mechanism works, it seems reasonable to suppose that Jupiter's radio emission must somehow be tied up with activity on the disk; possibly some features are more habitually noise-producing than others, in which case it should be possible to investigate the matter more closely. Unfortunately a radio telescope is a very inefficient instrument when it comes to pinpointing the source in the sky. A pair of binoculars can show details on the Moon only 20 miles across, but even the Jodrell Bank 250-foot telescope, one of the largest of its type in the world, cannot 'resolve' down to less than the Moon's apparent diameter. In other words, it established that *Lunik II* hit the Moon in September 1959, but it could not discern on which part of the disk the impact took place. In the same way it is known that Jupiter emits radio waves, but the tiny disk is far too small for selective analysis. The only way to link its emission with surface features is along indirect channels.

It was soon realized that the bursts occurred in a period corresponding roughly to the 9 hour 55 minute day of the planet's higher latitudes, thereby furnishing an obvious clue to the position of the source. The next step was to compare the radio results with ordinary visual observations made during the same period, and extensive use was made of the work by amateur astronomers belonging to the Jupiter Section of the British Astronomical Association. These indicated that the bursts tied in fairly well with transits of the Great Red Spot across the planet's meridian, as well as those of a number of nearby white spots. By 1964, however, the suggested relationship between the radio bursts and surface markings was being increasingly questioned, and further analysis provided the surprising conclusion that the bursts were related to the position of satellite Io in its orbit. The explanation is not yet known, but it seems certain that Io's movements are responsible.

Another interesting result of radio work is the discovery of a duplicate of the Earth's van Allen layers of electrons and other charged particles trapped in its magnetic field (page 167). Indeed, these results suggest that Jupiter's magnetism is considerably stronger than the Earth's. Saturn possesses a similar field, and very possibly the outer giants do as well.

To an observer with a small telescope the four bright satellites of Jupiter are just as interesting as the planet itself. Its total family comes to twelve, but of these eight are very small and dim. The main four are very bright indeed; they were discovered by Galileo in 1609 (hence the term Galilean satellites), and a pair of binoculars will show them at a glance. In fact they would be visible with the unaided eye were they split up and scattered in the night sky, but as it is they are masked by the overpowering brilliance of the planet itself. Details of the four Galilean satellites are given below.

Name	Mean Distance from Jupiter (miles)	Diameter (miles)	Orbital Period		
Io	262,000	2,000	1^d	18^h	28^m
Europa	417,000	1,750	3	13	14
Ganymede	666,000	3,000	7	3	43
Callisto	1,170,000	2,800	16	16	32

TABLE II – *The Galilean Satellites*

After a little practice they can be identified without the use of an almanac; Ganymede and Callisto show perceptible disks in quite a small telescope, Callisto having a curious purplish tint, while Europa, as befits its size, is appreciably fainter than Io. Surface markings can be made out with very large instruments, and observations carried out mainly at the Pic du Midi observatory indicate that they all keep the same hemisphere turned towards Jupiter. This is understandable if they have had a similar history to our own Moon.

These satellites are all oddly efficient at reflecting light. If the

Moon were substituted for Io it would appear much fainter, despite its similarity in size; the inference is that their surfaces cannot be so dull a substance as rock. The suggestion has been made that they are, like Jupiter, covered with frozen gases. Definite evidence is still lacking, but it seems a strong possibility; it would also explain why their densities are so low. Callisto, the least substantial of the four, has less than $1\frac{1}{2}$ times the mass of an equal volume of water, which means that any rocky core must be extremely small.

The movements of the satellites around their parent planet are fascinating to watch; Io and Europa, in particular, move so fast that a quarter of an hour will show clear displacement. The spectacle is enhanced by the fact that their orbits are exactly in the plane of Jupiter's equator, and since its axis is tilted a mere $3°$ from the vertical, we see the equator edge-on. Therefore the satellites appear strung out in a line, passing in front of the disk (transiting), and then swinging behind and being occulted. In just the same way as the New Moon, at the time of a total eclipse, casts a small shadow on the Earth, so the satellites in transit cast circular black shadows on the clouds. These shadows can be seen with a small telescope, and they may be more obvious than the satellite itself if it is seen projected against a bright part of the disk, and thereby partially camouflaged.

Jupiter's other eight satellites form a complete and remarkable contrast to the Galileans. In the first place they are all very small; so small, indeed, that their disks cannot be measured. Their sizes must therefore be inferred from their brightness, and assuming a reasonable reflectivity it is unlikely that any are more than 100 miles across.

The brightest, and closest, is Amalthea, which is only 70,000 miles above the clouds. At this distance Jupiter's attraction is tremendous, and Amalthea has to cover its orbit in less than 12 hours, travelling at a velocity of 1,000 miles per minute! This is not appreciably longer than Jupiter's day, so that its behaviour is rather similar to that of Mars' satellite Deimos.

The other seven moons have not been given names; instead

they are designated by Roman numerals signifying their order of discovery. Nevertheless an attempt is being made to attach logical deifications, and it is proposed to continue the effort here; it seems absurd that every minor planet should receive a name at the expense of Jupiter's family.

Number and Name	Mean Distance from Jupiter (miles)	Diameter (miles)	Orbital Period	
VI Hestia	7,124,000	80	250^d	16^h
X Demeter	7,192,000	25	254	5
VII Hera	7,302,000	12	260	1
XII Adrastea	13,000,000	12	600	
XI Pan	14,028,000	15	692½	
VIII Poseidon	14,620,000	25	739	
IX Hades	14,694,000	12	745	

TABLE III – *The Outer Satellites of Jupiter*
(The diameters are guesses only.)

This retinue is extraordinarily remote, and it can clearly be divided into two contingents: Hestia, Demeter, and Hera at the 7,000,000 mile mark, and the rest at double the distance. The outer four are so loosely held that they do not move in tight, defined orbits; the slightest external attraction, such as that of a passing asteriod, disturbs their motion, and their distances and periodic times can be regarded as only rough means.

The orbits of Hestia, Demeter, and Hera are tilted at 27° to the plane of Jupiter's equator. This is unusually large, but the outer four are even more extraordinary. We can envisage their orbits as having been spun through 180°, so that they circle Jupiter in the wrong direction.

The normal direction of motion in the solar system, viewed from the north side of the plane, is counter-clockwise; this includes both the rotation of the planets and their orbital travel, as well as the paths of their satellites, and of course strongly indicates a common origin. Apart from Uranus and its satellites, and the planet Venus, the only planetary objects to exhibit wrong-way or 'retrograde' motion are the four outer satellites of Jupiter and one member each in the families of Saturn and Neptune. Clearly, their origin must in some way have differed from the normal process, although their true significance has yet to be fully understood. It may be that Jupiter's outermost moons are nothing more than asteroids captured by its gravitational pull, even though the chances of this happening are extremely small.

The outer satellites would be of little use to a Jovian; assuming anything could be seen through the clouds, it would take a telescope to make them out at all! This means that they are extremely faint to terrestrial observers. Amalthea was in fact the last satellite to be discovered visually, in 1892, by E. E. Barnard with the 36-inch telescope of the Lick Observatory. The rest were discovered photographically, since the photographic plate is far more sensitive than the eye when it comes to picking up faint objects. In fact Adrastea is so dim that it has never been seen visually, even with the largest telescopes. Undoubtedly there are other still more fugitive moons waiting to be detected.

CHAPTER 10

Saturn

Mean Distance: 886,000,000 miles *Periodic Time:* 29½ years
Axial Rotation: $10^h\ 14^m$ *Equatorial Diameter:* 75,100 miles

UNTIL THE telescope came along the Ringed Planet was guardian of the solar system's frontier. But it earned no especial notice. Considerably fainter than Jupiter, it moved across the sky more slowly; its leaden hue earned it gloomy connotations in the profitable hocus-pocus of astrology. Saturn was left to its lonely destiny.

This banishment was not relieved until the discovery of Uranus in 1781, but Saturn itself started to present problems long before then. Galileo, fresh from his discovery of Jupiter's satellites, turned his telescope to the planet in July 1610. He saw a curious sight: a globe accompanied on either side by two smaller spheres that remained fixed in position from week to week. He told Kepler that 'Saturn consists of three stars in contact with one another', and announced in an anagram: 'I have observed that the most distant planet is triform.'

Puzzlement turned to astonishment eighteen months later, when the philosopher looked at Saturn again. The attendants had vanished, and he wondered if some 'mocking demon' was making fun of him. But no – they reappeared some months later, grew larger than before, and eventually developed into two great handles attached to both sides of the planet's disk.

The mystery was not cleared up until 1659, when Huyghens, using a telescope that was far superior to Galileo's primitive instrument, announced that Saturn was surrounded by a flat ring that nowhere touched the body of the planet. The explanation of Galileo's vanishing satellites is a simple one. His telescope was too poor to resolve the ring as such, and all it could do was show its main area as a sort of spherical blur. Its disappearance was caused by Saturn's axial tilt of 28°, which means that as it progresses around the Sun we see the ring system alternately from opposite sides, at one point passing precisely through its plane (Fig. 26). It is so thin that even in a large telescope it disappears for a day or two. This is what happened in the winter of 1611.

As a planet, Saturn's make-up is astonishingly similar to that of Jupiter, although it is obvious from the polar flattening

FIG. 26. *Different views of Saturn's ring*. This actually shows only half of the cycle (which altogether takes 29½ years), for after passing back through the edge-on position the opposite pole swings into view. Notice how both ring and planet cast shadows on each other, their extent depending on the relative position of the Sun.

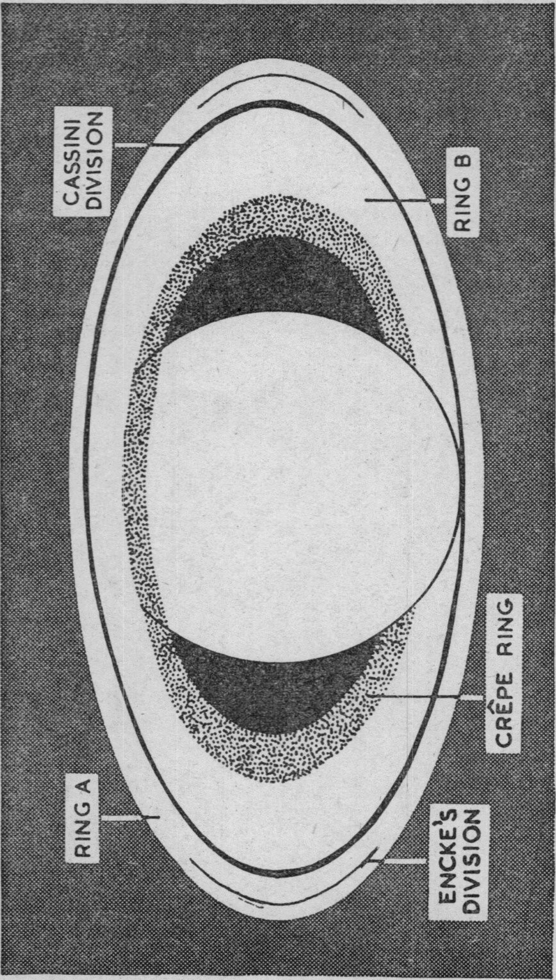

FIG. 27. *Saturn's ring system.* It is impossible here to indicate the different brightness of Rings A and B; moreover, the Crêpe Ring does not appear nearly so conspicuous as the diagram suggests.

that it is a less rigid body, a fact that follows from its unusually low density. Its polar diameter is only 67,200 miles, and because of the difference in gravitational effect any Saturnian in the gold trade could make a useful profit by buying at the equator and selling at the poles – provided he used a spring balance and provided also there were anyone else there to buy it!

Because of its remoteness Saturn appears much smaller than Jupiter, and markings are therefore harder to make out. The few that are seen, however, show that it also has an equatorial current with a much less sharply defined boundary; rotation slows with increasing latitude and is over half an hour longer by the time 60° is reached. It has its own equivalents of the north and south equatorial belts, but higher-latitude markings are fugitive. The general impression is of a less active world, and of course the temperature is slightly lower (—250° F). But occasionally patient observation is rewarded by the appearance of a white spot. These outbreaks are not frequent, however; there are only nine well-observed instances of white spots in the last two hundred years, and most of these have occurred near the equator. Recent spots include those of 1960 and 1962, but the most famous was that of 1933. This was discovered independently by two amateurs, one of them, Will Hay, being better known for his presence on stage and screen than behind a telescope; but Hay was an enthusiastic astronomer who had his own telescope and even wrote a short book on the subject. His spot lasted for several weeks, and it was a conspicuous feature during the early stages of its life. The spots of 1960 were at the unusually high latitude of 58° N, and a small spot seen in 1969, at 57° S, is the most southerly ever recorded on Saturn. Such high-latitude features are our only clue to the rotation period of these Saturnian regions.

Were it not for its ring system, Saturn would of course be a far less remarkable object; the gleaming oval, seen through even a modest instrument, is an unforgettable sight. There are actually four distinct rings forming the colossal annulus, whose total diameter is 169,000 miles. The outer belt is Ring A, 10,000 miles wide. Then comes a gap known as Cassini's Division, called after the observer who discovered it in 1675. This is 1,800 miles wide, and separates Ring A from the 16,500-mile Ring B. The next ring, Ring C (often known as the Crêpe Ring), is 10,000 miles wide and leaves a gap just wider than the diameter of the Earth around Saturn's equator. In this space, in October 1969, Pierre Guerin and his colleagues at the Pic du Midi Observatory photographed a new and exceedingly faint ring, Ring D. It appears as a ghostly band midway between the inner edge of the Crêpe Ring and the planet's equator, and has yet to be seen visually. A dusky ring outside Ring A, reported from time to time by various observers, has never been confirmed, and its absence from the remarkable Pic du Midi photographs seems conclusive.

These rings, which all consist of tiny fragments orbiting Saturn like individual satellites, differ greatly in brightness. Ring B is gleaming white, by comparison with which Ring A has a yellowish cast, while the Crêpe Ring is so dim that it was not discovered until 1850; it had been drawn before, but never recognized for what it was. These differences of intensity occur through the varying density of the particles, and the Crêpe Ring is so tenuous that when conditions are favourable the planet's globe can actually be seen through it.

This suggests that the rings must be thin, but practical investigation can be conducted only once every 15 years, when the Earth passes through their plane. When this happened in the autumn of 1789 the industrious Sir William Herschel gauged the thickness of the hair-line of light against the disks of the then known satellites. His result, using modern values for their diameters, came to about 250 miles. This thinness is remarkable enough, considering the colossal extent of the system, but recent investigations have reduced it still further: the present adopted value lies somewhere between 10 and 50 miles, and evidence inclines towards the lower value. Such attenuation is extraordinary. If we wanted to make a scale model, using thin typewriting paper, its width would be almost two feet! How did this precarious appendage come to be formed?

The original theory involved the so-called Roche Limit, which concerns the shattering of a satellite by its parent's tidal forces. It was advanced in 1848 by a young French professor, Edouard Roche, who pointed out that if a fluid or loosely-composed satellite approached its primary within a certain critical distance (the Roche Limit), it would be torn apart. The value for the Limit depends on a great many factors, but for a satellite similar in density to the parent it is nearly $2\frac{1}{2}$ times the planet's radius, measured from the centre of the planet. Roche therefore suggested that due to tidal influence an original inner satellite had slowly spiralled into Saturn's clutches and suffered this spectacular fate, just as Phobos, in Mars' family, is doing at the present time.

Unfortunately for this theory, the rings are certainly not liquid. They are composed of solid particles, together with a certain amount of ice, and a rocky, massive satellite would not have been disrupted in a manner suitable for the development of the rings we see today. Instead (and this is an amusing twist of scientific thought), modern ideas have completely reversed the sequence, suggesting that instead of being a shattered satellite the ring particles are the original fragments of a satellite that never coalesced. Very possibly the proximity of Saturn prevented these particles from accumulating, while the more distant regions of the primordial cloud gave birth to its other satellites in the normal way.

Collisions between the particles would gradually wear them down to roughly uniform size – they are probably on average a few inches across, although there is bound to be a considerable accumulation of dust as well. These collisions would also tend to flatten the system into a thin disk, so that the particles were moving in concentric orbits and collisions were reduced to a minimum. The thinner the ring, the more successful it would be; theoretically the ultimate result would be to have all the moonlets in the same plane. This gradual thinning-down is an inconceivably slow process, and we can never hope to gain direct evidence of it; what we see today is the virtual culmination of an infinite process.

A remarkable observation was made by Dollfus in 1958. By examining the light reflected by the rings he came to the conclusion that the particles must be cylindrical or cigar-shaped, pointing along their orbits – an ingenious piece of investigation.[1]

Another interesting observation makes use of the Doppler shift. By directing the spectroscope at the opposite radii of the rings it can be seen at once that the particles are approaching at the east side and receding at the west; they are therefore rotating in the same sense as the planet, which is something we should naturally expect. However, it also gives the death blow to the old 'solid-ring' theory by showing that the inner edge of Ring B is rotating faster than the outer edge of Ring A, the velocities being 11 and 9 miles per second. This is what we should expect, from Kepler's third law, if the rings consisted of independent particles. But if they were solid the outer regions would naturally have to travel faster than those closer to the planet.

The existence of Cassini's Division has been known for three centuries, but it took a long time to establish that it is a clear gap right through the ring. The most direct evidence comes when Saturn's orbital motion carries it in front of a star. The star disappears behind Ring A, but flashes out through the division; it is then swallowed up by Ring B, but manages to show dimly through the Crêpe Ring, if it is bright enough. There is also a gradual fading when it passes behind the ball of the planet itself, proving that the outer reaches of the disk are atmospheric.

Why should there be a division at all between the rings? There is an analogy here with the minor planets, in the way that Jupiter has produced gaps in certain zones that coincide with simple fractions of their revolution period. In the ring system's schisms the culprits are Saturn's satellites, particularly the nearby Mimas. A particle in Cassini's Division would have a revolution period equal to $\frac{1}{2}$ that of Mimas, as well as $\frac{1}{3}$ that of Enceladus, $\frac{1}{4}$ that of Tethys, and $\frac{1}{6}$ that of Dione – obviously impossible odds! The same rhythmic influence also applies to the much fainter Encke's Division in Ring A, which is not a true division at all but merely a thinning of the particles.

[1] A basically similar experiment to that performed by Lyot on Mercury's reflection.

Several others should theoretically be present, and many observers have claimed to see them, but so far they have not been fully confirmed. Cassini's Division corresponds to the gap near the great Hecuba group of minor planets, where the period is ½ that of Jupiter.

These divisions are probably of relatively recent formation, since they can have come into being only since the particles achieved fairly stable orbits; this means that we are seeing them in a transitive stage. Millions of years in the future, their aspect will probably be quite different; other divisions will have come into prominence and the rings themselves will have spread. A time-traveller would find here a fierce reminder that the universe never rests.

Saturn has ten known satellites. The family is more orderly than Jupiter's, and except for the outermost moon they form a regular group. Moreover they all have accepted names.

Name	Mean Distance from Saturn (miles)	Diameter (miles)	Orbital Period		
Janus	98,000	250?		17^h	59^m
Mimas	113,300	350		22	37
Enceladus	148,700	450	1^d	8	53
Tethys	183,200	750	1	21	18
Dione	234,600	900	2	17	41
Rhea	327,600	1,100	4	12	25
Titan	759,500	3,000	15	22	41
Hyperion	920,100	200	21	6	38
Iapetus	2,213,200	1,000	79	7	56
Phoebe	8,053,400	50	550½		

TABLE IV – *The Satellites of Saturn*
(The diameters are rough estimates only.)

Titan, discovered by Huyghens in 1655, is the brightest, and in a large telescope shows a perceptible disk; it is, however, very hard to measure accurately, although indications are that it is slightly larger than Neptune's Triton. In this case it is the largest satellite in the solar system. It is twice as massive as the Moon, and Kuiper has found indications of methane in its spectrum. This is the only definite evidence for any satellite having an atmosphere.

The diameters of the other moons, like those of Jupiter, are very uncertain. Iapetus is the most remarkable, for it is five times as bright when west of Saturn than when to the east. This may be due to its two hemispheres having unequal reflectivity, but it is hard to understand how the variation can be so great. It does, however, indicate that the satellite keeps the same face towards its primary, and so the same is presumably true of the other, closer, satellites as well. The Moon is certainly not alone in its irritating reverence.

The seven inner satellites move almost exactly in the plane of the rings and the equator, but Phoebe, a remote little world, revolves in a huge retrograde orbit that is inclined at an angle of 30°; the suggestion has once again been made that it is not an original satellite. It is interesting to reflect that here will be the best base from which to see the rings. The globe will be small – the same size as the Moon appears to us – but the rings, at maximum presentation, will be slightly more open than the 28° view we can get from the Earth. From Iapetus, whose orbit is inclined at 15°, the view is more oblique; and the closer satellites will show them almost edge-on. Even so, tiny Mimas must surely rate as the solar system's No. 1 beauty spot. The innermost satellite, Janus, was discovered in 1966 on photographs taken at the McDonald Observatory, Texas.

CHAPTER 11

Uranus

Mean Distance: 1,783,000,000 miles *Periodic Time:* 84 years
Axial Rotation: $10^h\ 49^m$ *Equatorial Diameter:* 29,300 miles

SOMEONE ONCE put forward an attractive though unlikely theory. Throughout our annual revolution around the Sun there is one point perpetually hidden from our eyes. This point is the opposite part of the Earth's orbit, which is always hidden by the Sun.[1] Could there not be another planet there, essentially similar to our own, but always invisible?

If a space probe today sent back evidence that such a world existed it would cause not much more sensation than Sir William Herschel's discovery of a new planet, Uranus, in 1781.

Herschel was an extraordinary man – no other astronomer has ever covered so vast a field of work – and since his name crops up again and again it is worth devoting a little space to his career. He was born in Hanover in 1738, left the German army in 1757, and arrived in England with no money but quite exceptional musical ability; he played the violin and oboe and wound up in Bath as organist in the Octagon Chapel. Herschel's was an active mind, and deep inside he was conscious that music was not his destiny; he therefore read widely in science and the arts, but not until 1772 did he come across a book on astronomy. He was then 35, but without hesitation he embarked on this new career, financing it by his professional work. He spent years mastering the then elementary art of telescope construction, and even by present-day standards his instruments are comparable with the best.

Serious observation began in 1774. He set himself the astonishing task of 'reviewing the heavens'; in other words, pointing

[1] Most unfortunately this is not strictly true, due to the eccentricity of the Earth's orbit and the resultant variation of orbital velocity.

his telescope to every accessible part of the sky and recording what he saw. The first review was made in 1775; the second, and most momentous, in 1780–1. It was during this that he discovered Uranus. Afterwards, supported by the royal grant in recognition of his work, he was able to devote himself entirely to astronomy. His final achievements spread from the Sun and Moon to remote galaxies (of which he discovered hundreds), and papers flooded from his pen until his death in 1822.

Among these was one transmitted to the Royal Society in 1781, entitled *An Account of a Comet*. In his own words:

> On Tuesday the 13th of March, between ten and eleven in the evening, while I was examining the small stars in the neighbourhood of H Geminorum, I perceived one that appeared visibly larger than the rest; being struck with its uncommon magnitude, I compared it to H Geminorum and the small star in the quartile between Auriga and Gemini, and finding it to be much larger than either of them, suspected it to be a comet.

Herschel's care was the hallmark of a great observer; he was not prepared to jump to any conclusions. Also, to be fair, the discovery of a new planet was the last thought in anybody's mind. But further observation by other astronomers besides Herschel revealed two curious facts. For a comet, it showed a remarkably sharp disk; furthermore it was moving so slowly that it must be a great distance from the Sun, and comets are only normally visible in the immediate vicinity of the Sun. As its orbit came to be worked out the truth dawned that it was a new planet far beyond Saturn's realm, and that the 'reviewer of the heavens' had stumbled across an unprecedented prize. Herschel wanted to call it after King George III (more than a trace of nepotism, perhaps), but the world rightly rebelled against this intrusion into mythological tradition, and Bode's suggestion of Uranus prevailed.

Uranus is a giant in construction, but not so much in size; its diameter compares unfavourably with that of Jupiter and

Saturn, though on the terrestrial scale it is still colossal. Its rapid spin has produced marked polar compression, but its disk is so small (equivalent in size to a halfpenny a mile away) that this is hard to make out in a small telescope. So are the markings; the only feature to be clearly recorded is a bright equatorial zone against the pale blue background. Uranus shines with a very curious blueness that is quite unlike any stellar tint, and it can be identified by colour alone.[1]

Recourse to the spectroscope shows that methane is present in great quantities, while there is hardly any ammonia. This is to be expected in view of the temperature (−300° F); only methane, hydrogen, and helium can remain gaseous under such conditions, and the two latter are hard to detect spectroscopically. It seems likely that the vast ammonia clouds we see on Jupiter and Saturn are absent, and that the atmosphere is relatively pure.

Detailed physical examination is out of the question, but considering the low temperature it would appear likely that the two outer giants are in a state of frozen complacency. However, Uranus occasionally shows curious fluctuations of brightness. It varies slightly in a rhythm corresponding to the rotation period – evidently due to dark or light features in one hemisphere – but there are also long-term variations over the years, which presumably signify periodic outbreaks of activity that affect its reflecting power. Despite these variations, however, the planet is always just visible with the naked eye. At the moment it is easily picked up in Virgo.

The most curious, indeed unique, feature of Uranus is the 98° tilt of its axis. No other planet exceeds 30° (we know nothing of Mercury, Venus, and Pluto), and this means that the Uranian seasons are unusual, to say the very least; the Earth's poles experience 6 months of perpetual sunlight, when the Sun grazes the horizon – but at the poles of Uranus the Sun reaches the zenith and stays above the horizon for 42 years! The corresponding polar night is equally long, and on the equator the 'annual' drift of the Sun takes it from the

[1] Many observers call Uranus greenish, but the writer (perhaps unreliably) sees it as blue.

northern to the southern horizon and back again. Travel agencies would certainly be in constant demand.

This means that on the disk seen from the Earth either the equator or a pole can be near the centre. In 1965 we had the normal equatorial view, while a pole will be presented in 1987.

Unanswerable proof that the origin of Uranus was closely connected with that of its five satellites follows from the fact that despite their parent's extraordinary tilt, they all revolve precisely in the plane of the equator. What is more, they form a remarkably orderly family.

Name	Mean Distance from Uranus (miles)	Diameter (miles)	Orbital Period		
Miranda	76,000	100	1^d	9^h	50^m
Ariel	119,200	400	2	12	29
Umbriel	166,100	300	4	3	28
Titania	272,500	600	8	16	56
Oberon	364,500	500	13	11	7

TABLE V – *The Satellites of Uranus*
(The diameters are guesses only.)

Herschel discovered the two brightest, Oberon and Titania, in 1787; he thought he saw more, but they turned out to be faint stars. An English amateur, William Lassell, found Umbriel and Ariel in 1851, and the fifth and faintest, Miranda, was photographed by Kuiper in 1948. It is so close to Uranus that it is lost in the planet's glare and is very difficult to see visually.

None of the satellites show disks, so that their diameters are once again mere guesswork; some estimates put the diameter of Titania as great as 1,500 miles. They have, in fact, earned relatively little attention, and the most interesting observations have been made by Dr W. H. Steavenson, using his 30-inch telescope at Cambridge. In 1926, and again in 1947, he noticed that both Titania and Oberon showed regular fluctuations of brightness. This is due, like Iapetus, to their hemispheres having unequal reflective powers, but there is also a more far-reaching inference to be drawn.

In 1926 we had an equatorial view of Uranus, and therefore

saw the satellites' orbits edge-on; by 1947 a pole was presented and the orbits had become almost perfect circles (Fig. 28). If the satellites followed the normal rule and revolved on

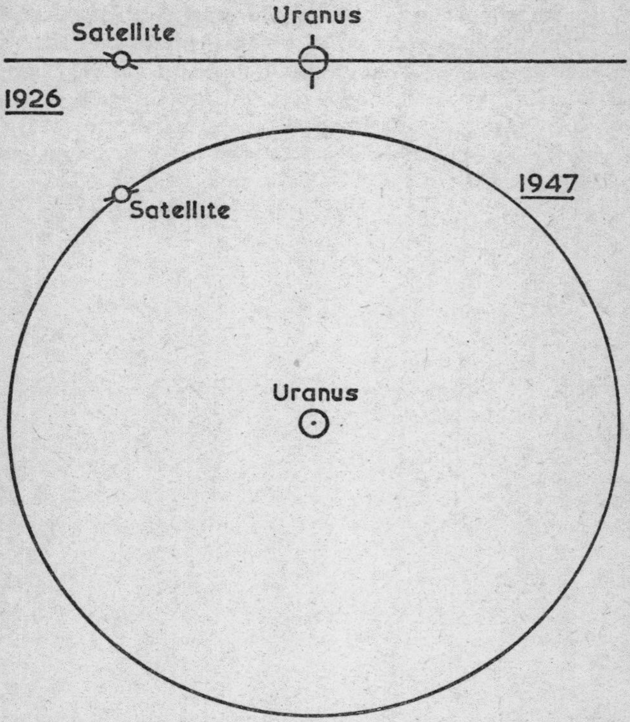

FIG. 28. *The moons of Uranus.* This diagram indicates why the behaviour of Titania and Oberon is so curious. A fictitious satellite is shown here.

approximately 'erect' axes (i.e. parallel to that of Uranus), their poles should also have been presented at this time and the same hemisphere kept continuously in view. The fact that the variations of brightness continued implies that this could not be the case. It therefore seems likely that Titania and Oberon resemble their parent in having their axes lying more or less in the plane of their orbits! It further seems reasonable that what is true for these two should also be the case for the others.

This is an astonishing situation. Our only certainty can be that Uranus itself was formed with this extraordinary tilt, as otherwise the satellites, condensing from its primordial cloud, would still revolve in orbits lying roughly in the solar system's general plane (they could not possibly tip up in sympathy). But why this should be the case, and why at least two of its moons should also show such a tilt, seems likely to remain an unsolved mystery.

CHAPTER 12

Neptune

Mean Distance: 2,793,000,000 miles *Periodic Time:* 165 years
Axial Rotation: $15^h\ 48^m$ *Equatorial Diameter:* 30,600 miles

HERSCHEL MAY have been the first to identify Uranus, but he was certainly not the first to detect it. It had been charted no less than 19 times beforehand, but always passed for an ordinary star; incredibly, one observer plotted it on four consecutive nights in 1769 but put the 'star's' motion down to errors of observation! These records dated back to the previous century, actually covering more than one complete circuit of the planet round the Zodiac, and were of course extremely useful in calculating the orbit. This made it even more remarkable that Uranus should refuse to obey predictions.

The basic theory was surely not at fault, and the observations themselves were reliable; even rejecting the pre-discovery positions (which might have been slightly inaccurate) made no difference. By 1845 Uranus was out of place by an angle equal to $\frac{1}{15}$ of the Moon's diameter; clearly an intolerable situation when planetary movements were habitually forecast to a hundredth of this error. Somewhere an unforeseen influence was at work, and it was possible that a still more remote planet was tugging at Uranus and pulling it from its true position.

This intriguing possibility attracted two gifted mathematicians, one English and the other French, who saw this as the only answer. Obviously the external planet was small and faint, as it would otherwise have been detected either by Herschel or (more likely) by von Zach's celestial police.[1] This meant that random searching was virtually useless. Somehow it was necessary to work out the intruder's position in the sky as accurately as possible so as to reduce the region to be

[1] This would have been a remarkable find indeed!

examined to practical limits. But could it be done? Most mathematicians considered the task impossible.

John Couch Adams, a brilliant student at St John's College, Cambridge, had no doubts about his ability to tackle the problem. In 1841, the year he entered the university, he made a note to investigate, when he had finished his studies, 'the irregularities in the motion of Uranus, which are as yet unaccounted for, in order to find whether they may be attributed to the action of an undiscovered planet beyond it; and, if possible, thence to determine the elements of its orbit approximately, which would lead probably to its discovery'.

He gained the Wranglership in 1843 (with more than twice the marks of the Second Wrangler), and immediately set to work. In less than two years he had reached his final solution, and called at Greenwich with the results. Unfortunately the Astronomer Royal, Airy, was away in France. Adams called again twice, but for various reasons Airy did not see him. However, he left a message stating his results, and Airy wrote to him asking for confirmation that his hypothetical planet explained a certain factor in the wanderings of Uranus. The answer was surely obvious enough, and Adams did not reply since he wanted to work through his calculations again – another year's work. Finally he sent his revised results to Airy in September 1846. But the Astronomer Royal was still reluctant to instigate a search.

Meanwhile, in France, Urbain Le Verrier had been working independently along the same lines. When he reached his own conclusions, in June 1846, he simply presented his results to the Académie Française, who duly published them. When Airy saw the paper he was astonished, for the two men had pinpointed the planet's position to within a degree. Doubt was superfluous; all that had to be done was look in the right spot. He detailed Challis, director of the Cambridge Observatory, to start the search. Unhappily, he could not have made a worse choice. Proceeding by the laborious method of charting the same region of the sky at intervals of a few nights, in the manner of the asteroid-hunters, Challis actually recorded Neptune on August 4th and 12th, but did not compare his observations. This is, perhaps, understandable. But, fantastic though it may seem, he actually remarked to his assistant that one of the stars seemed to show a disk, and might be the planet; on the following night, as he was going to the observatory to check up on it, he was side-tracked by the offer of a cup of tea, after which the sky had clouded up. The night after that was cloudless, but due to some trivial excuse he did not observe. Immediately afterwards came news that astronomers at the Berlin Observatory had found the planet, on September 23rd, by comparing the region of the sky with a new star-chart that had just been issued. Challis was not mistaken. He had seen Neptune, but had neither confirmed it nor mentioned his suspicion to anybody in authority.

The ensuing cross-Channel kerfuffle can be imagined, the French demanded Neptune for France, the English for England! Tempers flared, and Airy came under particularly heavy fire for having sat on Adams' results for so long. But the upshot was, after some extraordinary tactical blunders on both sides, to split the honours evenly between the two mathematicians who had dragged the new planet from its hiding place.

Early observations were discordant; several observers thought they had detected a ring around the planet similar to Saturn's. However, the same mythical ring had at one time been attributed to Uranus as well, and was less a tribute to keen sight than to telescopic imperfections. Neptune's tiny disk, only half the diameter of that of Uranus, is a severe test of definition. Even today we know almost nothing of the planet's physical condition. Methane is the only gas to show up in its spectrum, although hydrogen and helium are probably present; in the fierce cold (—330° F) there can be no gaseous ammonia. Undoubtedly we are once again seeing the top of a dense atmosphere.

The rotation period of Neptune, 15 hours 48 minutes, means that there are about 100,000 days in the planet's year! It is the densest of the four giants, although still much less substantial than the terrestrial planets, and it may have a small rocky core overlaid by a great thickness of ice. At the very least, it is an inhospitable world.

Neptune's gravest disservice to astronomy is its refusal to obey Bode's Law. The prediction is for a planet exterior to Uranus at 388 units, but Neptune's mean distance is only 301. However, as we shall see, Pluto fits Neptune's theoretical position fairly well. This raises obvious objections to Bode's sequence, and astronomers are still undecided whether or not it is due to pure coincidence.

Neptune has two satellites, and they are both remarkable bodies. The larger, Triton, was discovered less than three weeks after Neptune itself by Lassell, who was actually asked by Airy to join in the search with Challis but defaulted because of a sprained ankle. Triton is large as satellites go, probably only slightly smaller than Titan, and it probably possesses a methane atmosphere. At its distance of only 220,000 miles, rather less than that of the Moon, it covers its orbit in 5 days 21 hours. This is far shorter than the lunar month because Neptune's mass is 17 times that of the Earth, and it has to travel faster to remain in orbit. For the same reason the moons of lightweight Mars move relatively slowly.

Triton throws another spanner into the works of theoretical astronomy, for it revolves around Neptune in a retrograde direction. There is a significant difference between this massive, close satellite and the four tiny outer moons of Jupiter's system and Saturn's Phoebe. Because of this there is every reason to suppose that Triton's genesis was intimately connected with that of its parent; yet, if this is so, its motion seems inexplicable. It looks as though Neptune, as well as Uranus, may have had an unexpected history, and we shall return to the matter in the next chapter.

The second satellite, Nereid, was discovered by Kuiper in 1949, a year after Miranda. It is a small body, perhaps 200 miles across, but its orbit is an extraordinary one: its distance from Neptune varies from 867,000 miles at perigee to over 6,000,000 miles at apogee, in a period of 359 days. This is by far the most eccentric satellite orbit in the solar system. The outermost giant certainly has a remarkable family to accompany it on its wanderings through the lonely deeps of the solar system.

CHAPTER 13

Pluto

Mean Distance: 3,649,000,000 miles *Periodic Time:* 248 years
Axial Rotation: $6^d\ 9^h$ (?) *Equatorial Diameter:* 3,600 miles (?)

IN MANY ways the story of the detection of Pluto is the same as that of Neptune, though with a twist in its tail. Two mathematicians, this time both American, predicted the region of the sky in which it was eventually found. The astronomers were Lowell, of Martian fame, and W. H. Pickering, who held some odd views in the field of lunar and planetary astronomy but was nevertheless a first-class computer.

Slight remaining vagaries in the motion of Uranus were the cause of the inquiry. Any new outer planet would, of course, have considerably greater influence on Neptune; but its motion was less well known. Uranus had by this time been followed for well over one revolution (two if we include pre-discovery observations), while Neptune, at the turn of the century, had completed less than half an orbit. So in 1905 Lowell began to analyse the movements of Uranus, and by 1915 he had 'discovered' his Planet X. It was just over 4,000,000,000 miles from the Sun and had a period of 299 years. With a mass 7½ times that of the Earth, its diameter was evidently about 16,000 miles.

Curiously enough nobody followed Lowell's lead, and he died a year later. But in 1919 Pickering published his own results for Planet P. Its distance was 5,100,000,000 miles, its year 409 terrestrial years, and its mass only twice that of the Earth. Clearly planets X and P were incompatible!

Pickering instigated a search at Mount Wilson Observatory, and the observer in charge, Milton Humason, did in effect what Challis had tried to do sixty years before: he took photographs of the region indicated by Pickering at intervals of a few days, then compared the plates to see if any 'star' had moved. The planet's motion would be extremely slow, so that it would be easy to distinguish it from a minor planet – which in other ways it would of course resemble.

But Planet P refused to show up, and the search was called off. For a time the problem was in abeyance, until in 1928 the restless Pickering announced fresh results and urged another search. This was carried out, fittingly enough, at Lowell's own observatory, and the discovery of the faint, slow-moving planet by a now eminent astronomer, Clyde Tombaugh, was announced on March 13th, 1930, exactly 149 years after Herschel's discovery of Uranus.[1]

Why had Humason overlooked it? The subsequent inquest proved his failure to be due to nothing more than shattering bad luck. On the first plate Pluto's image had fallen on a flaw and did not show up, while on the second it was so close to a bright star that it was obscured by the glare. Nevertheless it was reasonably near to prediction, agreeing especially well with Lowell's elements.

Both Lowell and Pickering had predicted an eccentric orbit, and this transpired to be the case; for a few years around the time of perihelion the planet is closer to the Sun than Neptune, and this next occurs in 1989. There is, however, no chance of an actual collision, or even a particularly close approach, for the two orbits are inclined in different planes rather like two unequal hoops linked loosely together. In fact Pluto's orbit is tilted to the general plane of the solar system at an angle of 17°, which means that it can stray across a wide area of the sky and is not limited to the Zodiac. But its wanderings are very slow: at present it is in Leo, having managed only to crawl from the nearby constellation Gemini since its discovery.

The one surprise was its brightness, for it is considerably fainter than anticipated. To account for its observed effects on Uranus and Neptune, it should be rather more massive, and therefore presumably larger, than the Earth. This illusion was

[1] Moreover, it was within 15° of the discovery-position of Uranus.

somewhat disturbed in 1950, when Kuiper measured its minute disk with the 200-inch telescope and found a diameter of 3,600 miles. Obviously it will be asking a great deal of its density to fulfil Lowell's requirements; in fact, the matter composing it would have to be 50 times as dense as water or 9 times as dense as the Earth! This seems impossible. But how else are we to reconcile its discovery?

The feeling today is that the nearness of Pluto to Lowell's predicted place was coincidence, and various arguments have been put forward to explain how this apparently diminutive body could have the mass required of it by Lowell's analysis, none of them convincing. For instance, it was suggested that what Kuiper measured was not the whole disk of the planet, but merely a specular reflection of the Sun in a 'sea' of some reflective material on its surface. This idea, however, was disproved in 1965, when Pluto passed almost in front of a faint star without actually occulting it. Measurements proved that the planet must have a diameter no larger than 3,600 miles, or else an occultation would have taken place.

A further explanation put Pluto as the brightest member of an otherwise undetected minor planet belt beyond Neptune; while an indignant Pickering considered Planets P and X to be distinct entities, Pluto being Lowell's Planet X, while Planet P still awaited discovery! But a recent analysis of the motion of Neptune, using accurate positions up to 1968, seems to have cleared the matter up. R. L. Duncombe, W. J. Klepczynski, and P. K. Seidelmann, of the U.S. Naval Observatory, announced in 1969 that the mass of Pluto does not need to be nearly as large as Lowell thought. A mass of only 0·22 of that of the Earth accounts fully for the wanderings of Neptune, and this requires a density of 1·4 times that of the Earth. Although this still makes Pluto the densest body in the solar system, it is at least a plausible amount.

There is another oddity also. This is the length of Pluto's day, which light variations indicate to be roughly 6 days 9 hours long. This exceptionally slow spin requires an explanation. The only one so far to come to light, one which ties in the orbital peculiarity as well, is that Pluto was once a satellite of Neptune and somehow gave its parent the slip![1] There are obvious objections to this idea, but our total state of knowledge is so fragmentary that no suggestions can simply be dismissed out of hand.

Altogether, the outermost planet poses an uncomfortable number of problems: its discovery, its mass, its rotation, and its orbit. Is its good alignment with Bode's prediction for Neptune (mean distance 395) significant? On the whole it seems that it is not, for its orbit is extremely eccentric, perihelion distance being 2,766,000,000 miles, while at aphelion it recedes into murky twilight at 4,566,000,000 miles. And for adventurers, albeit in the imagination, it is exciting to speculate on the lonely worlds that may yet lie between us and the stars.

[1] Evidence for this comes from Triton's retrograde motion, which suggests that the birth of Neptune's family was a distinctly irregular event.

CHAPTER 14

Comets

IT TOOK the world a long time to find out much about comets. To the Greeks they were atmospheric phenomena, and the first practical hint that they are truly astronomical in nature came in 1577. The great comet of that year was observed by the intriguing tyrant Tycho Brahe,[1] who found its parallax to be less than that of the Moon, and its distance therefore correspondingly greater. To be fair, his were really the first instruments capable of such a feat.

Primitive people, some of whom still write today in newspapers and magazines, assigned to comets the rôle of celestial portents. Certainly a bright comet can be an awesome sight, but the remarkable truth is that they are all show and no integrity; the most brilliant comet that has ever flashed its sword across the sky is less massive than a minor planet such as Hidalgo, and by spreading this amount of matter through countless millions of cubic miles of interplanetary space, it becomes nothing more than an over-size ghost. On two known occasions, and doubtless on many in the past, the Earth has passed through a comet's tail and suffered about as much inconvenience as a lighthouse in a summer breeze.

A comet's basic collection of matter is in the nucleus. This is an accumulation of meteoric particles, probably laced with frozen gases, especially ammonia, not more than 500 miles across. Other gas, as well as meteoric dust and the like, forms a diffuse cloud or coma around the nucleus, and when a comet is still far away in the chill depths of space we can see it only as a faint blur of light.

Comets move around the Sun in orbits that are often extremely eccentric; at perihelion they may be as near as Mercury, but most recede at least to Jupiter's realm at

[1] His observatory, on the Danish island of Hven, included a dungeon, and among his retinue was a pet dwarf.

aphelion. As they drift closer to the Sun the increasing heat has a marked effect upon the coma. The gases vaporize, forming a great cloud which is forced away from the Sun by the pressure of its radiation (the 'solar wind'); as perihelion approaches the tail becomes more spectacular until, with a really great comet, it may stretch across a considerable arc of the heavens. After whirling round the Sun the comet is flung back into interplanetary space; the temperature drops rapidly and the tail disappears, and a few weeks later it may be invisible with the largest telescopes. The great thing to realize is that the tail, physically, is the least important item of the whole affair. It may extend for millions of miles away from the head, but nevertheless 99 per cent of the comet's mass is concentrated in the relatively tiny nucleus.

With a few spectacular exceptions, comets are all very much of a kind; it is their orbits that are of such great interest. They move around the Sun like planets, but in addition to their extreme eccentricity there is another important distinction: they show no tendency at all to adhere to the normal planetary plane. Some do, of course, but equally some have their orbits tilted at right angles, while others have, so to speak, turned right over and revolve in a retrograde direction. A comet may appear anywhere in the sky, from the celestial equator to one of the poles.

Comets may be conveniently divided into two groups. There are the regular or short-period objects, which return to the vicinity of the Sun again and again and can be predicted in advance, and there are others of much longer period (several centuries) that are normally observed only once. These are the really brilliant objects, but they are of less astronomical interest since our entire scrutiny may be limited to a few weeks.

The regular comets have periods ranging from $3\frac{1}{3}$ years to about a century, according to the length of their orbits, but a great many are grouped around the $6\frac{1}{2}$-year mark, and Fig. 29 shows the orbits of a few of the better-known ones; it will be noticed that their aphelia all occur in the region of Jupiter's orbit. The massive planet's attraction is responsible for this family, for comets, being so insubstantial, are easily

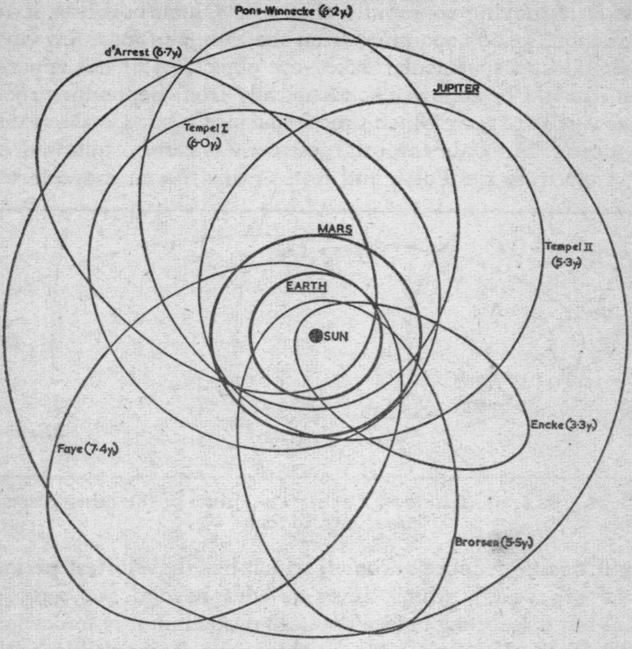

FIG. 29. *Jupiter's comet family*. For convenience Encke's Comet is also shown, although it does not belong to Jupiter's retinue.

dragged from their true paths. Oddly enough, however, none of the other giant planets possess so striking a retinue. The most famous comet of all, Halley's, has its aphelion well beyond the orbit of Neptune.

Halley's Comet has a period of roughly 76 years, and it has been recorded at almost every perihelion since 240 BC; it last appeared in 1910 and is due back again in 1986. Its orbit is shown in Fig. 30. At aphelion, which occurred in 1948, it was over 3,000,000,000 miles from the Sun. At the present moment it is moving back towards Neptune's realm, slowly gathering speed as it inches towards the Sun; by 1980 it will be closer than Uranus, and in 1985 it will pass Jupiter. After that it is moving so rapidly that a few months bring it to perihelion, 55,000,000 miles from the Sun, and for a few days it should be a spectacular naked-eye object. At its last apparition it could be followed telescopically from September 1909 until April 1911, by which time it had swept back to the realm of Jupiter, but only the brightest comets can be followed as far away from the Earth and Sun as this. As an example we

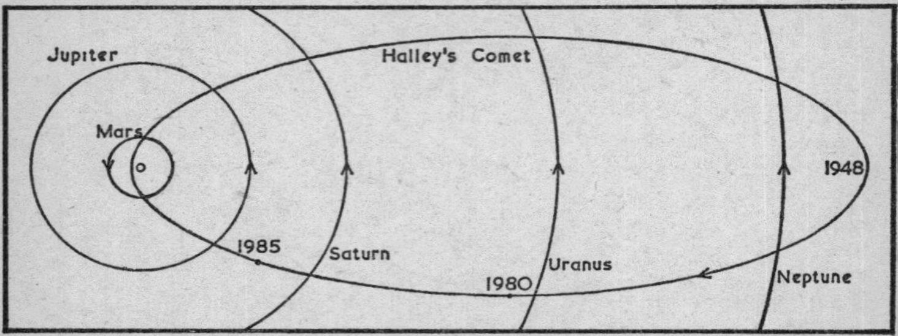

FIG. 30. *The orbit of Halley's Comet.* The orbits of the other planets are shown to scale.

might mention Encke's Comet, which has the shortest period of all ($3\frac{1}{3}$ years), which never recedes beyond 300,000,000 miles but is lost long before then. At perihelion it is somewhat closer than Mercury, and at this time it is visible with binoculars.

Encke's is an interesting comet. It was first seen in 1786, but not until the return of 1819 was its periodicity discovered; since then it has been observed at every return. Indications are that it once belonged to Jupiter's family and somehow slipped through the net. This evidence comes from a steady shortening of its period, amounting to two days per century – which is a rapid change on the astronomical time scale. This steady reduction may possibly be due to the comet encountering some resistance to its motion, possibly due to the incredibly tenuous dust which is known to litter space. It may seem ironical that slowing down its velocity should reduce the total period, but the explanation is simple. If any orbiting body loses speed, it is dragged closer to its primary; this is what happens to an artificial satellite on encountering atmospheric resistance. Encke's aphelion is therefore moving steadily towards the Sun; the total length of the orbit is being reduced, and as a result its periodic time is also lessening.

Another explanation, favoured by many astronomers, is that Encke's Comet is slowing itself down by internal explosions which act in a similar manner to the retro-rockets on a space craft. Some comets seem to undergo considerable self-

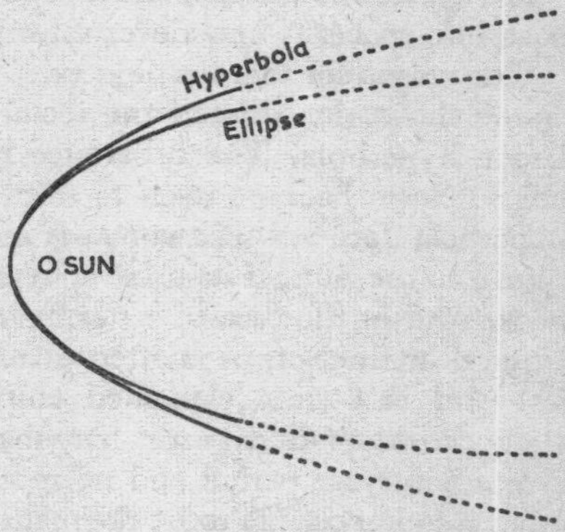

FIG. 31. *Basic comet orbits.* It is clear that the two types of path almost coincide in the vicinity of perihelion, which means that very accurate observation is necessary to decide whether a comet moves in a very elongated ellipse or a hyperbola.

destruction: Biela's Comet in 1854, Brooks' Comet in 1889, and Comet Ikeya-Seki in 1965 were all observed to break up into two or more components. It is therefore possible that Encke's Comet is slowing itself down, just as some other comets (such as Brooks') are accelerating in their orbits and increasing their periods.

About a hundred short-period comets are known, but Halley's is the only really bright one.[1] There is a reason for this. Spectroscopic observation shows that comets normally contain a great deal of dust, which is gradually lost by their bursts of splendour at perihelion; it therefore follows that the shorter the period, the more quickly this dust will be lost – and it is one of the basic ingredients of the tail. Therefore Encke's Comet, which has lost almost all its dust, can muster

[1] Halley's Comet is not normally counted as a short-period object, but it is the brightest of the regularly-returning ones.

hardly any appendage at all. Halley's has however been much more economical, and it still brings some of its pristine glory. Yet if we can trust the records, it is but a shadow of its former self, and obviously, if we wish for a truly great comet, we must seek a long-period object. This was brought home forcibly when a brilliant comet appeared in January 1910 and completely spoiled Halley's build-up. Utterly unexpected, it paraded its jealously-guarded glory for a few days, when it was visible in daylight, before shrinking and disappearing back into space. It may never return to the Sun.

The reason for this startling exit is shown by Fig. 31, which shows the region around the focus of two curves: an ellipse and a hyperbola. The difference between them is that an ellipse closes back on itself to form a complete ring, but the hyperbola does not. If a comet is moving in an ellipse, it will return to the Sun; if its orbit is hyperbolic, it can never come back. And in the case of a normal comet that is visible only when near the Sun, it is often extremely difficult to decide if its orbit is a very elongated ellipse or a hyperbola.[1] The slightest observational error may make the difference between a thousand-year period and no return at all, so that in these cases it is impossible to be dogmatic.

There is another factor also. Because of their small mass, comets are deflected from their true paths by the slightest influence. A close passage to a planet, especially Jupiter, can alter its orbit altogether, and a comet with a very long period may easily be flung into a hyperbolic path. Therefore although we may speak of a comet having a period of a hundred thousand years, it may well be lost altogether. The daylight comet of 1910 belongs to this hazardous class.

The brightest comets of recent years were those of 1965 (Ikeya-Seki) and 1970 (Bennett), both of which were best seen

[1] To be accurate, there are three basic types of cometary orbit, the third being of the parabolic class. The parabola marks the dividing line between the ellipse and the hyperbola, and if a comet is quoted as 'parabolic' it means that observations are inadequate to determine to which class it belongs. However, it simplifies matters to discuss only two classes.

by southern observers; but the most brilliant comet to appear in the last hundred years was seen in 1882. At the time of perihelion, in September, it was visible with the naked eye by simply blocking out the Sun with a hand, even though it was only 3° away from the disk! It had an unusually large nucleus, with a diameter of about 1,500 miles, and at greatest development the tail was detectable for nearly 100,000,000 miles. Ikeya-Seki was photographed on the day of perihelion (October 21st, 1965) when it was only ½° from the Sun's centre. The naked-eye comet of 1969, Tago-Sato-Kosaka, is of interest as being the first of these objects to have coverage from an earth satellite. Observations made from OAO-2 showed it to have a hydrogen 'corona' or halo, undetectable from the Earth's surface because its radiations were absorbed by the atmosphere, that was actually larger than the Sun! The launching of these 'observatories' is opening a whole new window into space.

Most comets are first detected several weeks before they reach perihelion, when they are still faint, nebulous objects against the background of the night sky. But as Fig. 32 shows, it is possible for a comet to sneak up in the region of the sky near the Sun, and so be completely invisible unless unusually bright.[1] This is what happened in the case of the comet of 1882, which was independently discovered by a number of observers just a few days before perihelion passage, which took it to within 300,000 miles of the photosphere. This meant that it shot round the Sun in just a few hours at about 300 miles per second, actually involved in the solar corona! If any proof were needed that comets are 'airy nothings' it is that this ordeal by fire apparently left it none the worse for wear.

An interesting thing happened as it receded into space: the nucleus divided into four parts which subsequently slowly separated along the orbit. It seems that in the far future these will return as separate comets, appearing about a century apart, the period of the shortest being some 650 years. This fact is suggestive: why should not other great comets of long

[1] During the total solar eclipses of 1882 and 1893 a comet was in each case seen close to the Sun, but it was observed neither before nor after the minutes of totality.

period have originally been constituents of a huge father comet, which for some reason divided? Evidence supporting this suggestion comes from a study of the orbit of the comet of

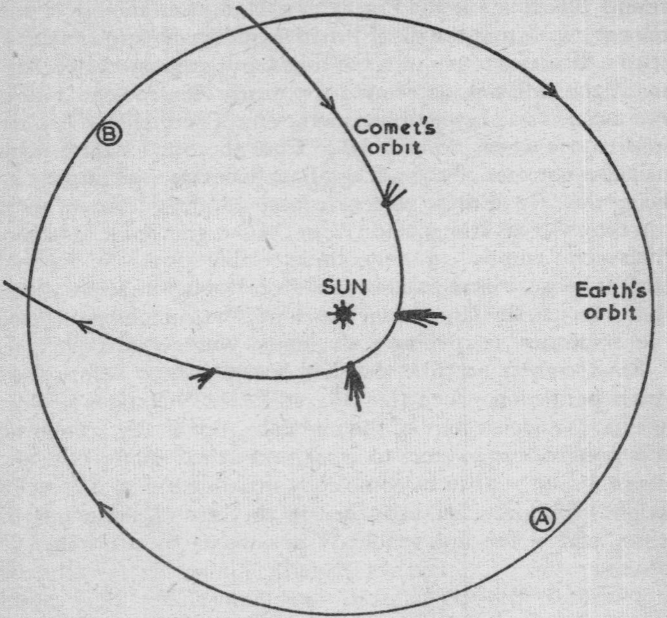

FIG. 32. '*Invisible*' *comets*. If a comet happens to come to perihelion when the Earth is in the region of A, it may appear too close to the Sun in the sky to be detectable. On the other hand, astronomers would have an excellent view if the Earth were at B. Thus the Earth dictates what comets are prominent just as much as the comet itself.

1882, comparing it with those of 1668, 1843, 1880, and 1887. In all cases the orbits are very similar, and it seems extremely likely that all these were once concentrated in what must truly have been a gigantic object.

Many comets are called after their discoverer, or co-discoverers in the case of independent detection. Sometimes there are other reasons also: Halley's Comet is so called because Edmond Halley, later Astronomer Royal, was the first to realize that the comet of 1682 was the same as that which had appeared in 1607 and 1531, and accordingly forecast another return in 1758. But there is also a more systematic method. When a comet is first detected, be it a discovery or a recovery of a known object, it is denoted by the year followed by a small letter signifying the order of detection. For example, the bright comet Arend-Roland was the eighth comet to be detected in 1956; it was therefore known as 1956*h*. There then follows the permanent designation, which gives the order of perihelion passage, denoted by Roman numerals. Thus 1956*h* was the third comet to pass perihelion in 1957, and its permanent name is 1957 III.

Mention of 1957 III brings us to the truly non-periodic or hyperbolic comets, which will never return to the Sun because they are moving too fast to remain in an elliptical orbit. An analogy can be made with the planet Pluto. If it were to suddenly speed up from its present 3 miles per second to the 4½ miles per second of Uranus, it would leave its normal orbit and fly off at a tangent. It would have reached what is known as 'hyperbolic velocity', and the Sun's attraction could never drag it back. The closer a planet or comet is to the Sun, the faster it must move in order to achieve this distinction; the Earth would have to speed up to 26 miles per second in order to escape.

Now in some circles there is fierce argument as to whether all comets are original members of the solar system, and the crux of the whole matter lies with the hyperbolic family. To help matters we can draw an analogy. Suppose that the Sun and some other particle were the only objects in the universe. No matter how far apart they were placed, the Sun's gravitational pull would eventually accelerate the particle towards its surface, and as it approached it would travel faster and faster, finally striking the photosphere at a velocity of almost 400 miles per second.

Suppose that due to some cosmic influence it just missed the Sun. What then? It would skirt the photosphere, rather in the manner of the comet of 1882, and soar back along a path almost parallel to the way it had come. The farther it sped into space the slower it would go. It would never return, but neither would it escape completely.

It is clearly impossible for a comet originating under these conditions to travel faster than this critical velocity. If we throw a ball up from the Earth's surface, it returns to the ground at precisely the velocity at which it was launched. But suppose someone up in a balloon caught the ball and hurled it back to the ground. In this case it would travel faster, since the balloonist has added his own velocity to it.

Imagine now that a star near the Sun has a planetary, and particularly a cometary, system similar to our own. If one of these comets were suitably perturbed by a planet, it would be thrown into a hyperbolic orbit and escape from the system. If it then chanced to pass near the solar system and be attracted towards the Sun, it would move with an acceleration due to the Sun's attraction (as did our original particle) *plus* whatever velocity it originally possessed (like the ball thrown by the balloonist). In other words it would be travelling faster than this certain critical speed; it would approach the Sun in a hyperbolic curve and after perihelion fly off again into space. Therefore if a comet could be observed approaching the Sun at hyperbolic velocity, we could be sure that it had been ejected from some other system in space.

Things are, unfortunately, greatly complicated by the presence of the planets. These have their own effect, which has to be weeded out, but the upshot is that no comets are known which move in hyperbolic orbits that are not brought about by planetary influence (the main culprit being Jupiter). Until we find a case of genuine hyperbolic velocity, we are safe in saying that all comets are original members of the solar system. This is not to say that other planetary systems do not also lose comets, but the chances of one passing near the Sun are almost negligibly small.

Furthermore, the solar system is losing comets at a steady rate – perhaps a dozen a year – and this drain has been going on for millions of years. Therefore theories of cometary evolution have to explain how, despite this colossal loss, there are still sufficient comets to keep astronomers busy. We must remember that only a small percentage of comets are bright enough to be noticeable, and the vast majority must slip by unseen.

There are two leading theories. Dr R. A. Lyttleton has suggested that they were formed from gaseous interstellar matter picked up by the Sun during its constant journeyings through space, while a more recent theory, advanced by Dr J. H. Oort of the Leiden Observatory, combines their genesis with that of the minor planets as due to the disruption of the original planet between the orbits of Mars and Jupiter. Oort points out that assuming the planet to have been similar in mass to the Earth it requires only a small fraction of the resultant debris to account for the existence of 200,000,000,000 comets, each with a mass of 10,000,000,000 tons! This mass sounds very large indeed, but on the planetary scale it is minute – equivalent to that of a world the size of Hermes. Of course, some comets contain much more matter than this.

The vast majority of these comets have been thrown by planetary perturbations into colossal orbits with periods of thousands of years, and it seems likely that they throng in cold, isolated loneliness in the dim reaches beyond Pluto's orbit. The theories of both Lyttleton and Oort successfully account for the vast number of comets that must exist, but one advantage of Lyttleton's theory is sometimes overlooked. This is that comets do not have to be attributed to one definite genesis; during its rotation around the centre of the Galaxy the Sun is periodically passing through interstellar clouds which replenish the supply of lost hyperbolic comets. Therefore the solar system will never run out of its orbiting ghosts, whereas Oort places a definite time-limit. However, it is so far ahead that the Sun itself may have died before its initial store runs out.

It is interesting to reminisce on some of the more remarkable visitors to the Sun's vicinity. Early observers were perhaps over-enthusiastic; in chronicling the comet of 146 BC as being 'as bright as the Sun', fear had presumably secured a firm grip on scientific accuracy. Yet many ancient records are remarkably accurate, containing sufficient observations of great comets to enable rough orbits to be computed.

It is difficult to decide which comet of the last four centuries was really the brightest. The contest is presumably between those of 1577, 1744, 1811, 1843, and 1882, and certainly nothing has so far turned up this century to touch any of these. Yet apparent brightness is really a very unfair measure if we wish to judge their true worth; a near comet will outshine one of equal brilliance a greater distance away. Therefore if we want to find the 'greatest' comet of recent times we must turn to a relatively dim object that appeared in 1729 – it was just visible with the naked eye when at perihelion. But whereas the perihelia of the above comets were all less than a million miles from the Sun, that of the comet of 1729 was about 370,000,000 miles away – almost as remote as Jupiter! When the great comet of 1882 had receded to this distance it was barely discernible without a very powerful telescope, so that had it approached to a normal perihelion distance the 1729 colossus would undoubtedly have swept the rest off the board altogether.

Marks for aesthetic appeal usually go to de Chéseaux's Comet of 1744, which appeared in the morning sky as a huge six-tailed fan rising before the Sun. Another notable visitor was Donati's Comet of 1858. Some comets have been remarkable for exhibiting rapid changes in their constitution; perhaps the most notorious is Morehouse's Comet of 1908, which shed great clouds of material from its coma in just a few hours. Coming to more recent times, Comet Arend-Roland exhibited a distinct spike or 'beard' of luminous matter pointing directly towards the Sun, which seems to be a comparatively rare feature.

There are two reasons why a comet brightens on approaching perihelion. The first is simply increased reflection of sunlight; the second is actual emission of light from the gases produced by the decomposing particles. The tail, in particular, shines almost entirely by its own light. Analysis of its spectrum usually reveals evidence of water, nitrogen, carbon dioxide, ammonia, and methane, whose molecules are split up into other compounds by the high temperature. One of these is the lethal gas cyanogen (CHN), and when it was announced that on May 19th, 1910, the Earth was due to pass through the outer extremity of the tail of Halley's Comet, scientific alarmists had a field day. However, comets are so incredibly rarefied (it was calculated that a cubic inch of ordinary air contained more material than 2,000 cubic miles of the tail!) that no harm could possibly have come to our innocent population; in fact, there was no visible sign that we were in the tail at all. During a previous event of similar nature, on June 30th, 1861, the Earth passed through the tail of the bright comet of that year and there was a perceptible glow in the night sky – but once again no physical effects were noticed.

These encounters, of course, raised the inevitable question: what would happen if a comet's nucleus struck the Earth? The answer is, a great deal; but luckily cometary nuclei are so small, compared with the vast volumes of their tails, that the likelihood is about as great as that of being struck on the head by a meteor during a country ramble.

CHAPTER 15

Meteors and Meteorites

To date there is no well-documented case of anyone having suffered a direct hit by a meteor. This is perhaps hardly surprising, for probably no more than a few thousand meteoric bodies strike inhabited land each year. But these are the cream, and the very rarefied cream, of the meteoric crop; these are the few which survive the total fall through the atmosphere. The rest, millions of them, perish daily in a streak of fire, their vaporized remains adding to the mass of the Earth at the rate of about 2,000,000 tons per year.

The scientific world did not become thoroughly meteor-conscious until the nineteenth century; although they were accepted as atmospheric phenomena, their real astronomical connexion was not finally proved until the memorable night of November 12th, 1833. During the early hours of the morning the United States witnessed an extraordinary celestial firework display. Shooting-stars flashed across the sky by the thousand, some stations counting, or at least estimating, 200,000 meteors between midnight and dawn. Nothing like it had ever been seen before, and the world had to wait until 1966 for a repeat performance. During the shower a number of white-hot stones fell; satisfactory proof that meteors and meteorites were basically one and the same. And meteorites, of course, had been known for centuries.

Meteors are simply extremely small particles of matter circling the Sun in the manner of the planets. Some of them travel in isolation, but most are gathered together in great clouds or streams which usually extend along the entire orbit. If this orbit should chance to intersect the Earth's at a certain point our planet will then encounter a rapid and momentary increase in meteoric activity, something that is known as a meteor shower.

Meteors travel at speeds, relative to the atmosphere, ranging from about 10 to 45 miles per second, depending from what angle they approach the Earth (since the Earth's orbital velocity of 18½ miles per second will clearly have a considerable effect). First atmospheric encounter occurs at an altitude of about 120 miles, but not until they fall to 70 miles does the fierce resistance offered by the air cause them to incandesce. Normally their fate is sealed in about a second, burnout occurring at 50 miles, but objects larger than a grain of sand will penetrate lower and glow more brightly – anything bigger than a pebble will produce a streak bright enough to illuminate the entire sky. A really brilliant meteor like this is usually known as a fireball, while one that explodes in flight, with or without an audible explosion, is a bolide.

The vast majority of meteors are mere specks of dust, and leave so faint and fleeting a trail that they are quite invisible with the naked eye; on an average night, under good conditions, a watcher will see only about six meteors per hour. These are known as 'sporadic' meteors, which means that they have no connexion with any shower. But when the Earth passes through a swarm the hourly rate increases. The best showers are those which reach maximum intensity on August 12th and December 13th (although some outliers of the swarms are visible for several weeks on either side of these dates). It is obvious that since the orbits of both the Earth and the meteors are fixed in space, encounter must occur at the same time each year.

When a shower occurs the meteor trails are noted and plotted on a star chart, the stars affording very convenient reference points. It is then seen that by back-tracking the trails they can be made to meet in a certain small area of the sky, a region known as the radiant. This does not mean, as it may seem at first sight, that the meteors are really radiating from a point in space; it is entirely an effect of perspective. They are travelling around the Sun in parallel paths, but when we see them flying towards us the effect is analogous to that of standing on a railway track; the rails appear to converge on a point at the horizon. Furthermore, the meteors during any particular shower must appear to come from the same general direction all the time. A certain effect of parallax does occur due to the Earth's orbital motion through the stream, but generally speaking the radiant remains in the same constellation for the duration of the shower.

This is very convenient, for the shower can be named after the constellation in which its radiant is situated. The August meteors appear to fly from a point in Perseus, and are therefore known as the Perseids; the December shower (radiating from Gemini) are the Geminids. Altogether there are some 30 identifiable showers throughout the year, with many minor ones, which means that at certain times two may overlap. A meteor seen on August 1st may be a Perseid or an Aquarid, or it could of course be sporadic.

The intensity of sporadic meteors suffers a diurnal variation; on average roughly twice as many are seen at 6 am as at 6 pm, supposing that conditions are dark. This is because before sunrise we are on the Earth's leading hemisphere in its race around the Sun, and our own orbital velocity scoops up a large number of meteors; also their speed is effectively increased and they are therefore brighter. In the evening, however, they have to do much more work to catch up with us, and only the fastest succeed. This does not apply to proper showers, for the meteors are in these cases all travelling in the same direction, and only the hemisphere facing the swarm will receive a proper display.

Since meteors are genuine members of the solar system, it is obviously of the greatest interest to work out the orbits of the various showers. In themselves they are of course far too tiny to be visible except when plummeting through the atmosphere, and research is therefore confined to analysis of their trails; the problem is not to simply observe their apparent path across the stars, but to find the real direction of flight through the atmosphere. To do this two or more observers are required, stationed about 50 miles apart and watching the same region of the sky. Since meteors flash into view around the 70-mile mark there will be a clear parallactic effect (Fig. 33), and simple trigonometry enables the meteor's path to be calculated. By studying large numbers of samples from the same shower, deductions can be made about the behaviour of the swarm as a whole.

In this way it has been found that meteor swarms move in orbits that are remarkably cometary in form – even though there is no evidence of hyperbolic paths, thus disproving a

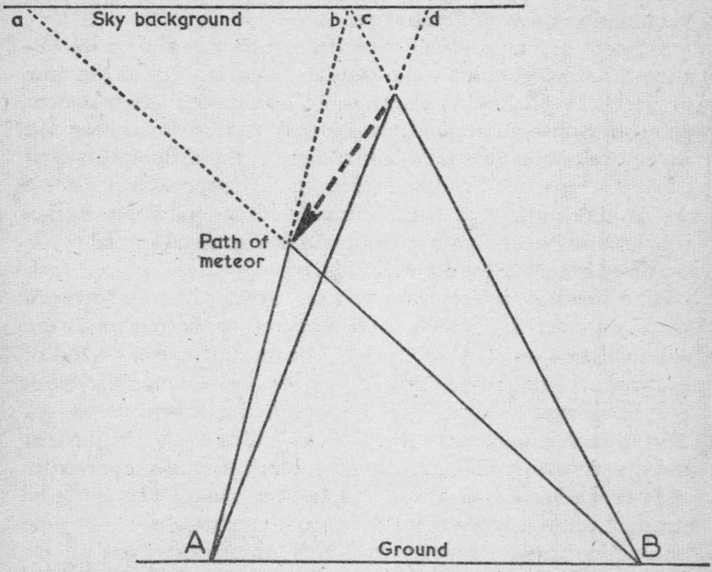

FIG. 33. *Calculating a meteor's path.* Observer A sees the meteor's path across the sky as *bd*; to observer B it covers the much larger arc *ac*. Combination of these values leads to knowledge of the meteor's true path through the atmosphere.

belief held by some until recently that they came from interstellar space. The most remarkable discovery of all is that some meteor swarms move in the orbits of known comets. Actually this is not new; it dates from the suspicions of naked-eye observers working at the turn of the century, and there was also the curious episode of Biela's Comet.

Biela, an Austrian astronomer, discovered his comet in 1826, and, like Halley, found it to be identical with an object observed previously; with a period of $6\frac{3}{4}$ years it obviously belonged to Jupiter's family. At the return of 1845, to the great astonishment of observers, it performed a slow amoebic division into two parts; by 1852 the two comets were several million miles apart. In 1859 the comet was too badly placed to be seen at all, but in 1866, when it should have been well visible, there was no trace of it.

Comets are notoriously unreliable objects, and most observers had written it off as a bad job when in 1872, at the time of the next phantom return, there occurred a great meteor shower. Subsequent analysis showed that the meteors did indeed follow the comet's orbit; somehow, in its spectacular suicide, the comet had scattered its meteoric debris to the winds. Thereafter the shower became an annual affair, occurring around November 27th, but with the gradual spread of the particles it is now hardly identifiable.

The Leonids, which produced the extraordinary shower of 1833, were active again in 1866; apparently the meteors were confined to a small section of the orbit, and had a period of 33 years. It then transpired that their orbit coincided with that of Tempel's Comet, which naturally had the same period. The inference once again is obvious, and it is significant that the comet was missed at its 1932 return, being recovered in 1965 only after a special search had been made. The expected strong Leonid shower of 1899 failed to materialize, but 1966 brought a brilliant display to North America, where, on the night of November 16/17, rates of up to 40 meteors a second were recorded. This was calculated to be twice as intense as the original shower of 1833. The nearby attendance of Comet Tempel cannot just be regarded as coincidence; undoubtedly, the Earth passed through a dense patch of meteors recently ejected from the nucleus.

Many other cometary associations, some of them admittedly dubious, have been made. It is hardly surprising that they fail to account for the total number of showers, for within the last century at least two comets have disintegrated and left legacies. Yet these showers have rapidly become very feeble, and it seems likely that the great and long-lived showers, such as the Perseids (which have been recorded for at least a thousand years), had a quite independent origin – possibly connected with the formation of the minor planets.

Meteor astronomy is one field in which instrumental advances have had decisive effects. Naked-eye observation was never very satisfactory, since it is exceedingly difficult to memorize accurately the beginning and end of a trail that lasts for half a second, and the few prolific observers were all lone workers. The greatest of all meteor observers, W. F. Denning (1848–1931), lived in Bristol – from where, incidentally, he discovered three comets.

In more recent times, with photographic emulsions becoming ever more sensitive, accurate observation has been handed over to the meteor camera. These, while recording the trails very precisely, do however suffer from one drawback: they are less acute than the eye, and miss faint meteors altogether. There is therefore still a certain amount of work for the visual observer to do in simply noting the frequency with which meteors appear during a shower. At maximum the Perseids and Geminids usually muster about 60 meteors per hour, but other showers are considerably less prolific. On the other hand, unexpected things sometimes happen; in 1956 amateur observers in South Africa discovered a new and remarkable shower on December 5th, now known as the Phoenicids, which are unfortunately only visible in the southern hemisphere.

Much the most interesting development in the technique of meteor observation has been the introduction of radar. When a meteor hurtles through the atmosphere it has an effect on the surrounding gases (mainly nitrogen) known as ionization: the intense heat causes their atomic structure temporarily to change, and this is actually what causes the streak of light. This transient line of ionized air can be observed by radar, and very accurate values for velocity and direction are possible.

Radar investigations were first carried out at Jodrell Bank in 1946, and their blessed independence from weather and daylight soon brought the remarkable discovery of a great summer meteor shower lasting from May until August, with a maximum in the middle of June. Due to sheer bad luck these meteors appear to be coming from the direction of the Sun, which means that they always fall on the daylight hemisphere and are therefore invisible. This is just one of the many important advances made by radar, and it is certainly the most powerful tool in the observer's arsenal.

Analysis of the physical composition of meteors is a very important branch of astronomical research, since meteors might be considered as the building-blocks of planets; the substances in them are presumably to be found in planetary cores as well. There are two ways in which this can be done. The most obvious is to analyse meteorites, which are rather larger meteoric bodies that have managed to survive the fall through the atmosphere. Unfortunately, these large bodies are not really representative of the meteors that move in showers; they are sporadic, and there are very few instances of meteorites actually landing during a meteor shower. Some people, in fact, consider them as tiny minor planets in their own right. Showers are made up of much smaller bodies that burn out completely in the atmosphere, and the only way in which these can be studied is by photographing their spectra.

This is obviously a very difficult task. Only bright meteors have a chance of leaving their spectrum on a photographic plate, and there is no way of telling whereabouts in the sky they are most likely to appear; it is mostly a question of luck, and dozens of exposures may have to be made for each spectrum caught. To date the world total is about 300 meteors. The main elements betrayed by this means are iron, nickel, calcium, manganese, and magnesium, which agrees well with the chemical analysis of meteorites.

The dominating material is usually either iron or calcium, and this is reflected in meteorites themselves, which can be divided into three classes. The majority of those found are essentially iron, and are known as siderites. Another smaller class consists predominantly of stone (calcium), and these are called aerolites. Finally there are the siderolites, fringe objects that fall between the two divisions. There is actually no reason to suppose that siderites are more numerous than aerolites, but casual searchers are far more likely to identify the former, since aerolites can easily be mistaken for ordinary stones. Siderites are usually curiously pitted by uneven burning in the atmosphere, and are therefore more conspicuous.

Why should there be these distinct classes? The 'disrupted planet' theory offers an explanation, since if an Earth-like planet were responsible for the Sun's family of minor planets, meteors, and comets, we should expect the proportion of iron particles to stone to be roughly what it is. What is more, all meteorites show a crystalline structure that can only have come about through fierce heating and subsequent cooling, and since such tiny bodies could not themselves have acquired such a temperature they must subsequently have been fractured from the main mass. It may well be that the minor planets exhibit the same classes; Vesta, for example, could be a metallic mass that has remained unusually bright. This is only a suggestion, but it is at least possible.

Most museums have a few small meteorites on display, and the largest is the 33-ton object found in Greenland by Peary in 1897. It is almost 11 feet long and is now exhibited in the Hayden Planetarium in New York. The largest meteorite in the world, however, is still in its resting-place near Grootfontein in South-West Africa; it weighs about 60 tons. By contrast, the heaviest meteorite ever found in the British Isles is the Limerick stone, weighing 106 lb, which fell in 1813. This meteorite may, however, have been exceeded in weight by the object which fell in the Leicestershire village of Barwell on Christmas Eve, 1965. The original meteorite disintegrated in flight, scattering 23 major pieces in fields and gardens – one landed in a flower pot, and another hit a parked car. Put together, 103 lb of meteoritic material was gathered, and undoubtedly many pieces have never been found. Before this, the English record was held by the 64 lb stone which fell at Hatford, Berkshire, in 1628.

However, the Earth bears the scars of far vaster impacts in the past. The greatest of these is the 7-mile Chubb crater in Ungava, Canada, which has become a great circular lake; the depth from floor to rim is about 1,300 feet. More spectacular,

PLATE I. *A Complex Sunspot.* This spot group, about 200,000 miles long, was photographed by amateur astronomer Ramon Lane on October 26th, 1969. Note the clear difference between umbra and penumbra, the mottled 'granulation' of the solar surface, and the penumbral 'bridges' across the principal spot.

PLATE II. *The Solar Corona.* Total eclipse of the Sun, taken by H. C. Hunt at Yurgamysh, Siberia, on September 22nd, 1968. A 3-inch refractor was used, exposure $\frac{3}{4}$ second on FP4. The irradiation of the prominences causes notches in the outline of the Moon. Visually, they appeared red against the white corona.

PLATE III. *Comet Bennett, 1970.* This photograph of one of the brightest comets of the century was taken by D. S. Brown, an amateur astronomer, with a home-constructed telescope, on April 27th, 1970. Note the double tail and streamers.

PLATE IV. *Nova Persei, 1901.* These two photographs, separated in time by about 35 years, show how the cloud of matter resulting from the star's outburst has expanded. It would take millions of hydrogen bombs to simulate such a disaster.

PLATE V. *Three close galaxies.* The beautiful spiral M.81 (*a*) is a mere 7,000,000 light-years away, which is why we can see it in such great detail. NGC 4594, the 'Sombrero Hat' galaxy (*b*), we see almost edge-on, and the absorbing dust in the plane of the arms is very distinct. It is this dust, in our own galaxy, which prevents us from seeing the nucleus. The Greater Magellanic Cloud (*c*) is almost formless, though there are indications that it is beginning to develop arms and turn into a spiral. It probably represents the 'youngest' type of stellar system.

PLATE VI. *The Pleiades.* The five bright stars are easily visible with the naked eye, but the curdling nebulosity is hard to detect visually, even with large instruments. This residual gas proves the youth of the system, about 410 light-years away, and indicates the superiority of photography in such fields. Photograph by W. E. Pennell with a 10-inch reflector on February 13th, 1971.

PLATE VII. *The Orion Nebula.* This huge cloud of glowing hydrogen is 1,500 light-years away. Note the dark clouds partly obscuring our view of the bright nucleus, which is made to shine by excitation from hot blue stars in its midst. Stars are probably being formed in this 'nursery' at the rate of several per century. Photograph by W. E. Pennell with a 10-inch reflector on December 7th, 1970.

PLATE V.

THREE CLOSE GALAXIES

(a) M.81

(b) NGC 4594

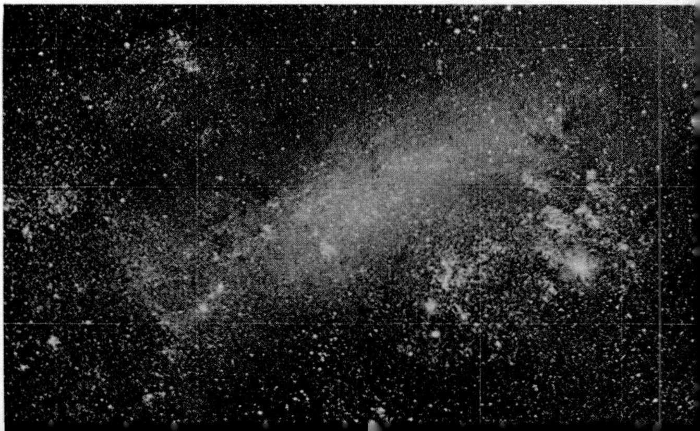

(c)
THE GREATER MAGELLANIC CLOUD

PLATE VI. THE PLEIADES. The five bright stars are easily visible with the naked eye, but the curdling nebulosity is hard to detect visually, even with large instruments. This residual gas proves the youth of the system, about 410 light-years away, and indicates the superiority of photography in such fields. Photograph by W. E. Pennell with a 10-inch reflector on February 13th, 1971

PLATE VII. THE ORION NEBULA. This huge cloud of glowing hydrogen is 1,500 light-years away. Note the dark clouds partly obscuring our view of the bright nucleus, which is made to shine by excitation from hot blue stars in its midst. Stars are probably being formed in this 'nursery' at the rate of several per century. Photograph by W. E. Pennell with a 10-inch reflector on December 7th, 1970

however, is the dry Barringer crater in Arizona, 4,150 feet across and over 500 feet deep. There are no records of the meteoric falls that produced these vast legacies, which, considering their sites, is hardly surprising, but erosion studies suggest that the Barringer crater is between 5,000 and 10,000 years old. Many efforts have been made to excavate the meteorite concerned, but with no success, and since meteoritic fragments have been found all over the nearby countryside it seems probable that it exploded on impact. In its original state it must have weighed many thousands of tons, with a diameter of perhaps thirty or forty feet. This small size may at first sight seem surprising, but what digs a meteor crater is not the physical size of the body so much as its sheer energy of motion, as well as the intense local compression of the air.

By far the greatest fall within living memory occurred in the Yenisei Valley, Siberia, on June 30th, 1908, and mutilated the local reindeer population – although, miraculously, no human lives were lost. The roar was heard hundreds of miles away, and several people saw the trail despite daylight conditions, but the site was not visited until twenty years later; by that time the marshy region where it fell had swallowed up the crater, and all that could be found were a few meteoritic fragments. There was another fall, less severe, in south-east Russia in 1947. So far these monsters have avoided built-up areas (statistically the chances of one scoring a direct hit on a city are about one in 100,000 years), but they clearly cannot land in the Soviet Union indefinitely, and there is always the very remote chance of a disaster – sufficient at least to keep the alarmists happy.

Meteorites are our only tangible link with outer space, and the most fascinating cosmic problem is the question of life in the universe: just how common is it? Mars beckons with its dying deposits of lichen, but what of elsewhere? Is the living cell as fundamental a unit as the ninety-two natural elements, or is it a mere chance arrangement of chemicals?

Two American scientists, Dr George Claus and Professor Bartholomew Nagy, have recently been examining meteoritic samples for signs of living organisms. The problem is not an easy one; quite apart from the difficulty of identification, how do we define a 'living' cell? The best that can be done is to compare what is found in the meteorite, if anything, with life-forms found on the Earth.

Only one type of meteorite seems suitable for such a study, and this is a sub-class of the aerolite family which contains a good deal of carbon; this is because the carbon atom is the building-block of terrestrial living matter. Such 'carbonaceous' meteorites are very rare, the total known weight being only a few pounds, and the two meteorites examined by Claus and Nagy were museum specimens: one fell in Orgueil, southern France, in 1864, the other in Ivuna, central Africa, in 1938. In 1961 they managed to isolate a number of carbon compounds closely resembling animal fats, while a few months later they discovered five varieties of what they call 'organized elements'. At sight, four closely resembled aquatic algae, while the fifth seemed to be entirely new. Altogether about 2,000 of these minute bodies were recovered, and further research by other workers on the same meteorites has confirmed these finds.

It is not suggested, of course, that these are literally living cells; for one thing they are millions of years old, and the severe conditions to which they have been subjected must have fossilized them in the same way as the petrified trees found in Jurassic forests. This naturally makes the process of identification much harder, and work is still continuing. There are many sceptics who consider them to be purely mineral in nature, possibly produced by the effects of cosmic rays during the aeons for which the bodies have circled the Sun, and as yet this possibility has not been disproved – but we know so little about cosmic radiation that it is little more than a shot in the dark.[1]

[1] Recently a damper has been thrown on this research by the work of W. Fitch and E. Anders, of the University of Chicago, who have found the 'entirely new' organized elements to be identical with a species of ragweed pollen(!). It therefore seems likely that the meteorites' contents were, after all, due to terrestrial contamination.

If it *is* proved that they are the remains of living cells, which presumably came to fruition on the primeval planet, there are startling inferences to be drawn. First, carbon-based life of the terrestrial variety is very likely universal (thereby disposing of Plutonians made of solid plutonium!); second, the formation of fundamental living cells is a common phenomenon. This is a fact which would carry staggering implications for the theologian and biologist alike.*

CHAPTER 16

The Earth's Surroundings

THE SKY is never really dark. Contrary to popular ideas, there is still a little light left (quite apart from starlight) even on a moonless night at the top of a mountain. To escape from this final excessively tenuous shroud we should have to rise 600 miles above the Earth's surface. Such an altitude takes us well into the astronomical realm, but it happens also to be the upper boundary of the auroral zone.

A brilliant display of the Northern Lights can be an exciting experience, but this permanent faint aurora gives no pleasure to the astronomer. Forcing him to look through its screen of radiation, and thereby dimming very faint objects beyond, it is one more hint that the days of terrestrial-based observation are numbered; that further progress in many departments of optical astronomy must await rocket-borne telescopes and cameras that can fly clear of this haze and see to the limits of the universe. Such refinement is not yet possible, but space probes have at least helped to give us a better understanding of the aurorae and of the Earth's environment in general.

There is no mystery about an aurora. If we take a strong glass tube and pump out the air until the pressure is reduced to 1/1,000th of its normal value, and then pass a current between metal electrodes sealed into both ends of the tube, the rarefied gas will glow. If, instead of pumping out the air, we filled the tube with air at an altitude of 100 miles or more, we should get the same result. When the molecules, particularly those of nitrogen (which makes up 78 per cent of our atmosphere), are very dilute, they give off light under the influence of an electric current. An electric current is nothing more than a stream of electrons, and each electron disturbs the electrical balance of a nitrogen atom, causing it to emit a minute amount of energy. This is what produces the glow.

It has been known for many years that auroral displays are

closely linked with solar activity. They fluctuate in a period corresponding to the sunspot cycle, and almost invariably occur when a large sunspot, particularly if associated with a flare, has just passed the Sun's meridian. Since flares are known to emit intense radiation, including electrons, it was naturally supposed that on reaching the vicinity of the Earth the electrons curved

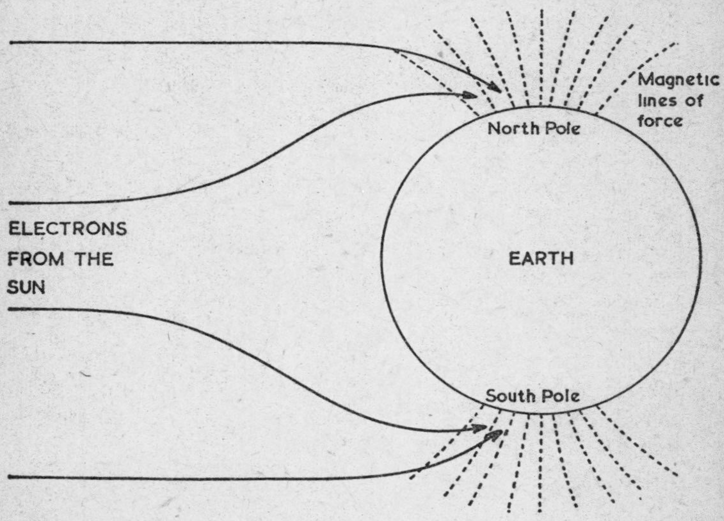

FIG. 34. *Simple theory of the aurora.* This idea was prevalent before the discovery of the van Allen zones.

down to meet the atmosphere along the lines of magnetic force (Fig. 34). They would therefore influence especially those atmospheric regions in the vicinity of the poles, which explains why aurorae rarely occur in places near the equator.

This was very simple and satisfactory until *Explorer I* started its task of, among other things, examining the radiation intensity at heights ranging from 230 miles to 1,600 miles. Much of its time was spent inside the normal auroral zone, which naturally contained a good deal of radiation; but above 600 miles, instead of fading out as everyone expected, the measurements

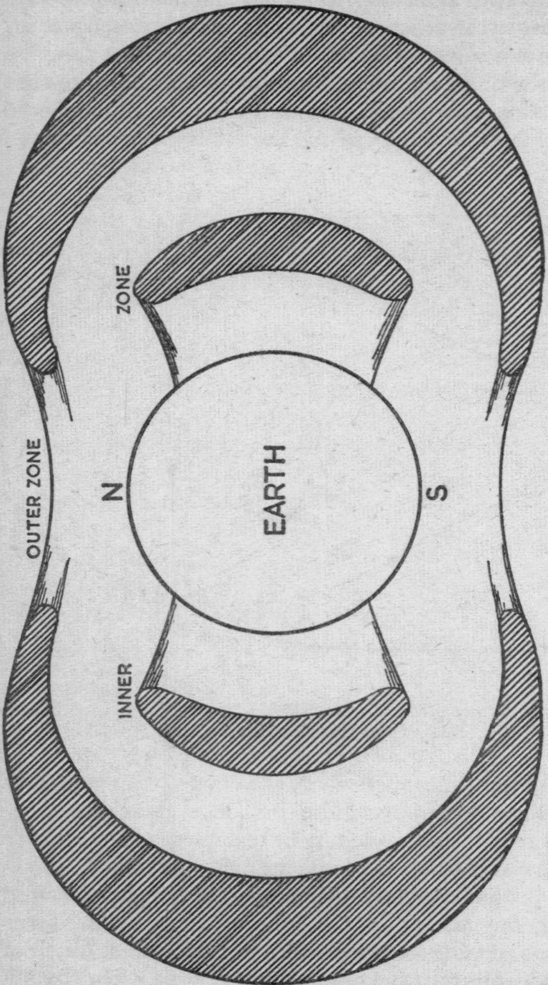

FIG. 35. *The v* *Allen zones.* T represents cross-secti through t whole syste The zones are, course, far mo diffuse than shown here, a their actual e tent varies wi solar activity

shot up again. Later probes, among them *Lunik II*, not only confirmed this layer of permanent radiation at an altitude of about 2,000 miles, but also discovered another much more intense layer at around the 10,000-mile mark. These are shown in Fig. 35. They are known as the van Allen zones, for it was Dr van Allen, of the Iowa University, who was responsible for installing the initial equipment in *Explorer I*. We now know that the Earth's radiation belts extend out to about 40,000 miles. These remote regions, however, are not symmetrical; as Fig. 35 shows, they have been deformed into an aerofoil contour by the force of the solar wind. However, the van Allen zones mark the most intense areas of radiation, and they are clearly important to both the astronomer and the astronaut. They consist of enormous numbers of charged particles trapped in the Earth's magnetic field, and their make-up is very strongly contrasted: the inner zone consists almost entirely of positive particles (protons), while the outer zone is rich in negative particles (electrons). What is more, this zone is so dense that its radiation is extremely heavy – well above the danger limit for human life if exposed to it for prolonged periods. It is, however, very unstable. A burst of solar radiation can make the entire layer vanish in a matter of seconds, and it takes several hours for the everlasting solar wind to refresh it with a new supply of electrons. Should the radiation hazard prove a really serious one, it would always be possible to delay a launching until a suitable flare neutralized the belt!

The existence of this layer raises a serious objection to the 'classical' theory of aurora formation; since electrons are trapped in the zone, they clearly cannot influence the high-altitude nitrogen atoms. Apparently what happens is this. The short-wave radiation which is part of the burst has a serious effect on the Earth's magnetic field, releasing the electrons from their trap and allowing them to flow down to the atmosphere. With the door opened, so to speak, further direct emission can reach the atmosphere, only to be blocked out again when the magnetic field stabilizes itself and the layer re-forms.

Since any powerfully magnetic body must trap protons and

electrons in a similar way, there is no reason why other planets should not also possess similar radiation layers. Venus and Mars do not seem to have fields comparable with that of the Earth, but Jupiter and Saturn certainly do, and they apparently are both encased by the equivalent of our van Allen zones.

Auroral displays can take a great many forms, ranging from a greenish glow near the northern horizon to a medley of arcs and beams which riddle the sky like searchlights. Sometimes these beams are parallel and give the illusion of a great curtain hanging in the heavens, while on some occasions they ascend from various points on the horizon and appear to converge at the overhead point (the zenith), giving the impression of a huge canopy. Green, red, and white are the predominant colours in an auroral display.

A faint aurora, which looks for all the world like the beginning of the dawn twilight arc – except that it is in the north – may be no brighter than the Milky Way; a really brilliant one may almost rival a Full Moon, and there are rare cases of aurorae having been seen in broad daylight. This happened twice recently, in 1957 and 1958, and of course it is not without significance that the recent sunspot maximum in December 1957 was the most active ever recorded. The greatest nocturnal display to be seen from England in recent years was that of January 25th, 1938, when the red beams could be seen even from places in southern Europe. This means that activity is occurring very far from the normal limits, and knowledge of these so-called 'tropical' aurorae, as opposed to the usual polar type, is very incomplete. Polar aurorae are distinguished by the term 'borealis' or 'australis' according to whether they occur at the north or south zone.

It is obvious that high-latitude observers have the best chance of seeing auroral displays. To dwellers in the far north of Scotland they are the rule rather than the exception around the time of sunspot maximum, but the south of England is never likely to witness more than eight or ten even in a very favourable year, and most of these will be visible only to the patient watcher.

It may be argued that aurorae are hardly astronomical; but

if this is so, then neither are meteors. Both are atmospheric phenomena triggered by some extra-terrestrial impulse, whether it be a tiny grain of matter or a sudden surge of electrons from the outer van Allen layer. But there can be no doubt at all of the interplanetary nature of two other nocturnal illuminations which are much more persistent than aurorae. In fact they are always present, but they are so obscure that few people have ever seen them. They are the Zodiacal Light and the Counterglow, often referred to by the German equivalent of Gegenschein.

To explain the Zodiacal Light we must go back to Chapter 1, where the young Sun was surrounded by the cloud from which its family of planets was later to form. In the first instance this cloud was in the form of an irregular mass, slowly rotating. This rotation quickly evened out irregularities; more than this, it flattened the cloud from a sphere into an increasingly squashed ellipsoid, so that by the time planet-sized aggregates had started to form it had become a disk just a few million miles thick. Once the accretion processes were under way, the gravitational attraction of the proto-planets quickly cleared the cloud of virtually all the large lumps of matter. But they left behind an extensive haze of small particles, ranging from dust to meteoric bodies a few inches in diameter. Occasionally one of these bodies lands as a meteorite, but they are so widely scattered (just one in hundreds of cubic miles of space) that collisions do not often occur. It also means that despite the vast depth through which they are ranged, they obscure a negligible amount of light from the stars beyond. This, from the astronomer's point of view, is very fortunate; the interplanetary fog is extensive but very thin indeed.

Everyone has seen the effect of a shaft of sunlight shining into an otherwise dark room: its beam is a swirling mass of dust specks that are otherwise invisible. We get the same effect when we look at the Sun through this residual fog; sunlight shining on the particles creates a meteoric aureole around the Sun. Now this halo is extremely weak, so weak that we cannot normally see it even during a total eclipse because of the brightness of the corona. In practice, of course, it merges with the corona, since the particles exist right up to the vicinity of the Sun, and their reflection of the light reinforces the brightness of the solar atmosphere itself. However that part of the glow which is truly interplanetary – more than a few million miles from the Sun – is doused.

Since the meteoric bodies are in the form of a disk, we can expect the halo to be extended in the plane of the solar system. Therefore there should be a possibility of seeing its outer ramifications either after sunset or before sunrise, when the Sun itself is well below the horizon and the sky is still dark. And this is actually what happens. On a very clear September morning or March evening a pale cone of light can be seen slanting up into the sky, and this is the outer extension of the halo, known as the Zodiacal Light. At its base the Light is about 30° wide, and it tapers to an apex about 70° from the Sun's position. Near the base its brightness is rather greater than the most vivid parts of the Milky Way, but of course atmospheric absorption near the horizon reduces this greatly, and at its edge and apex it is very faint indeed.

It is therefore hardly surprising that the Light should be so intermittently seen, even though it is a permanent feature of the sky. However, British observers do not have the best of it. In temperate latitudes it never rises very high in the sky, whereas in the tropics, where the Sun rises and sets vertically, the cone appears upright and is much more easily seen – in fact it is a quite normal sight.

The fact that the Zodiacal Light is an infant of the Sun only by proxy is an irritating one, and several observers have made attempts to actually see it during a total solar eclipse, thereby demonstrating its connexion with the corona. As well as solving the difficulty of the overpowering brightness of the inner corona, however, there is also atmospheric diffusion to add an extra veil across our view of the Sun! One of the earliest constructive observations was made by Professor Langley in 1878, who took the trouble to climb Pike's Peak, in the Rocky Mountains, and observe the eclipse of that year from a height of 14,000 feet. The benefits he obtained were considerable; the corona extended to a distance of about 6° on either side of the eclipsed Sun, an angle equivalent to about 10,000,000 miles! Obviously this was the root of the Zodiacal Light proper. It is also interesting to record that the sky was so transparent that he could see the inner corona for some minutes after the brilliant photosphere had reappeared. It is but a step to conditions on the airless Moon, where the corona and its zodiacal ramifications are a permanent feature of the black sky, and no eclipse by the Earth is needed to show them.

Night sky illumination does not end at the apex of the Zodiacal Light. Under first-class conditions a band of light, considerably fainter than the cones, can be seen encircling the star sphere, running along the Zodiac. This occurs for the simple reason that interplanetary bodies also exist beyond the Earth's orbit, although the light they reflect back is less than that received from the particles closer to the Sun and therefore in a more oblique position. Hence the so-called Zodiacal Band is very faint indeed, and only a few observers have ever seen it. It is about 10° across, and where it crosses the Milky Way it is overpowered and lost from sight. The Band effectively joins the apexes of the two cones, and with the Sun as the nucleus of the whole affair it is not unlike living at the centre of a diamond ring.

At the centre of the Band, exactly opposite the Sun in the sky, is an ellipse of light measuring roughly 10° by 7°: the Counterglow. This reinforcement of light in the Band is to be expected for optical reasons, and although brighter than its background it is still excessively faint – far dimmer than the Milky Way.

Regular observers of these strange phantoms have noticed that on some nights, apparently equally transparent as on other more favourable occasions, they have been entirely absent. This is not to suggest that they are in themselves variable. What probably happens is that for some reason the tenuous aurora that illuminates the upper atmosphere brightens sufficiently to douse these other glows, even though the sky itself still appears perfectly dark. Needless to say, perfect atmospheric conditions are required for a view of such vague features as the Band and the Counterglow, and only a few people have seen them.

PART TWO

Stars and Galaxies

The homely solar system must now be displaced by the far vaster issues of the universe. Our Sun is but one member of the Galaxy, a spiral system comprising 100,000,000,000 stars; and great telescopes can detect millions of galaxies. Astronomers' probes into these regions are all the time working towards the ultimate riddle: How did it all start?

CHAPTER 17

The Night Sky

SO FAR we have talked only of the nearby bodies that move across the face of the sky in front of the stars – though for all the use our eyes are in estimating relative distances, we might as well say 'among' the stars. For who is to prove us wrong if we say that Mars is moving 'through' Leo? More important still, who is to deny that it is much more convenient to consider the Moon and planets as creeping across the inner surface of an imaginary body known as the celestial sphere, to which the stars are attached and which rotates round the Earth once a day?

The ancient astronomers probably never gave this consideration much thought, since it was obvious that the celestial sphere *did* exist; how else could the Sun-god manage his daily journey from east to west? They were less concerned with finding out about the stars themselves than with documenting the sky as they saw it, by sorting out the star patterns and judging their various brightnesses. It was in this way that the backbone of observational astronomy, the constellations, came into existence, and we must spare some thought for their makers whose judgement gave us what we now see as the constellations; whose patient observation tells us what the sky was like thousands of years ago.

The origin of the constellations is usually attributed to the Babylonian–Sumerian peoples of roughly 3000 BC, whose imagination drew figures among the stars and gave us the very ancient constellations of Aquila (the Eagle), Gemini (the Twins), Hydra (the Sea Serpent), and one or two others. Whether or not this is so, we can at least be sure that all the sky accessible from 30° N and higher latitudes had been documented by the time the Greek thinkers appeared on the intellectual scene in about 600 BC. From our point of view the most important Greek was Hipparchus (190–120 BC), who might be called the first practical astronomer – an antediluvian Herschel. Hipparchus compiled a catalogue of 1,080 stars divided among 48 constellations, and in doing so he laid the foundation of the present notion of stellar magnitude, one of the most fundamental units of astronomy.

It is obvious that some stars are far brighter than others, and Hipparchus differentiated between them by dividing them into six different classes or 'magnitudes'. The most brilliant he described as of the 1st magnitude, while the faintest visible with the naked eye were of the 6th magnitude.[1] The rest occupied the units in between. Therefore his catalogue was one not only of position but also of brightness; it was the first attempt at a truly scientific description of the heavens, and as such it is of tremendous value.

These six magnitudes, in modified form, provide the basis of modern brightness classification. Telescopic observation has naturally extended the range of observable stars, and while the faintest visible with the eye are still roughly 6th magnitude, the 200-inch telescope can photograph stars as faint as the 23rd magnitude! Also, the steps now correspond to a standard ratio, so that a difference of five magnitudes corresponds to a brightness ratio of 100. In other words, it would take 100 6th magnitude stars to equal the brightness of one 1st magnitude star. A difference of one magnitude means a ratio of roughly $2\frac{1}{2}$.

At the upper end of the scale revisions have also had to be made, for accurate measurements have shown that some stars are brighter than the 1st magnitude; there is therefore a zero

[1] Notice that the higher the number, the fainter the star.

magnitude, while three stars actually have negative values. Sirius, the brightest star in the sky, has a magnitude of −1·44, while the Sun and the Full Moon are −26·7 and −12·8 respectively.

It is clear that these measurements are representative of the stars only as they appear to us; if we are looking along a street at night the distant lamps look fainter than the close ones, whereas we know that they are of the same real brightness, or luminosity. Stars do not all have the same luminosity, but the principle is the same, and Sirius appears the brightest mainly because it is relatively close. We therefore define their brightness in the sky as their 'apparent' magnitude. A star's luminosity, or real brightness, is called its 'absolute' magnitude, and to find this requires special investigation.

The problem of apparent magnitudes was settled many years ago, but the constellations themselves posed difficulties of their own. The shepherd-astronomers had not been too precise in defining exactly what part of the sky belonged to what constellation, and the addition of more modern figures only served to confuse the matter still further. It would be a pity to abolish them, since this artificial grouping together of stars forms an easy way of mapping out the night sky; on the other hand it is essential to know where the stars belong, and in the case of the dimmest naked-eye objects there was often considerable difference between the catalogues. Obviously official boundaries were needed. Matters were not finally put to rights until 1930, when the International Astronomical Union dictated the precise regions occupied by all 89 constellations. This ended the interminable struggle by the adjacent constellations Auriga and Taurus for a bright star that the IAU finally adjudged the property of Taurus!

Most of the naked-eye stars are named according to the constellation in which they lie, and there are two systems in common use. The first, which confines itself to the brightest stars, was devised by Bayer for his map of 1603: he attached a Greek letter to each star, followed by the genitive of the constellation name, and generally speaking the letters were in the order of magnitude. For instance, the brightest star in Lyra is called α

Lyrae, while the second and third are β and γ. This is an excellent system – or would have been, had Bayer been systematic. But unfortunately his order frequently lapsed when he came to the fainter members, while he abandoned his system with constellations such as Ursa Major, where he labelled the seven brightest stars in order of position, along the line. Nevertheless the system is still in general use, and it does form at least a rough guide to the likely magnitude of a star.

Bayer's system caters for only 24 stars, so it is likely to run dry in a large constellation. For the fainter objects we therefore turn to the catalogue issued by Flamsteed, the first Astronomer Royal, in 1725. Flamsteed observed the positions of virtually all the naked-eye stars visible from Greenwich, and the total number in his catalogue comes to 2,923 – a tremendous piece of work. These stars he assigned to their constellations by a number applied not in order of magnitude but in order of position, from west to east. Flamsteed included Bayer's original stars in his lists, but in these cases the Greek letters are usually used in preference.

There have of course been many other catalogues issued since Flamsteed's time, but on the whole these are used for specialized work only and do not take constellations into account at all.

Some stars have been given names. These are mostly of Arabic origin, and although several hundred have been handed down only a handful are still in general use: α Lyrae is usually referred to as Vega, while α Canis Majoris is Sirius. There are some real mouthfuls in existence, such as Alkalurops, Kornephoros, and Zuben el Genubi, which have not surprisingly fallen out of use. The same may also be thankfully said of various southern constellations 'invented' during the last two centuries. The Chaldeans had of course been unable to document the sky right up to the south celestial pole, and this was not done until recent times; hence some of the southern constellations are rather banal compared with their mythological northern compatriots. There is a Clock, a Compass, and also a rather incongruous Air Pump, but these are positively celestial in comparison with Officiana Typographica and Spectrum

Brandenburgicum, which together with other cumbersome relations have left the heavens for good.

The broad face of the sky, which has hardly changed since old Hipparchus compiled his immortal catalogue, is therefore a fascinating kaleidoscope of history. Some may find it disconcerting to have to deal with galactic red-shifts one minute and Greek heroes the next, but it is an ever-present and invaluable reminder that astronomy is truly the most ancient of all sciences.

CHAPTER 18

The Stars

OURS IS a very insular view of the universe. Asked what is the most important object in the night sky, the temptation is to answer: the Moon. Encouraged to go farther afield, the planets Venus and Mars might struggle on to the rota. This, after all, is reasonable enough; it would matter very little to our political and social ambitions were every star suddenly blotted out, but if the Moon were spirited away overnight there would be a distinct air of frustration in the astronautical camps.

Therefore turning from examination of the Earth and its immediate environment to the vast and terrifying reaches of interstellar space demands not one but two adjustments of the mind. The first is to increase the scale of everything a thousandfold. The second, and much more difficult, is to revise our personal sense of proportion. In the world of stars there is no room for planets, simply because they are so far away that no telescope could possibly detect them. If we were to transport the 200-inch telescope to the vicinity of the nearest star, the Sun would appear as naked as all the other stellar points of light. An inhabitant of these regions could never know that this yellowish star was the centre of a system of nine planets, one of which boasted a reasonably intelligent civilization. The probability is growing that as many as 50 per cent of 'normal' stars (i.e. those resembling the Sun) possess planetary systems, but we have not the remotest chance of ever being able to see one of these worlds for ourselves.[1] On the stellar scale planets are reduced, relatively, to the rôle of interplanetary dust; we cannot see both elephants and microbes with the same instrument.

Even the nearest star, which is a faint object in the southern constellation of Centaurus, appears only as a point of light; no star shows a disk, even in the greatest telescopes. This is a tribute to their distances, for many, in fact most, are actually

[1] Although we can in fact detect their presence (page 200).

larger than the Sun. Stellar observation is therefore principally concerned with analysis of their spectra, which tells us about their composition and gives clues to their physical characteristics, and accurate observation of their movements so that we can establish their paths through space.

The Sun is our prototype star; it is the only one we can examine in detail, and indications are that it represents very broadly the majority of stars. Their mechanisms for producing radiation are basically similar, depending on the conversion of hydrogen into helium, and because of this they all consist principally of hydrogen. This can be read in their spectra, which also of course tell the story of the other elements present as well – and this is of such vital importance that it should be dealt with first.

Stars are divided into eleven different classes. These classes are denoted by letters, and the sequence, in order of descending surface temperature, runs W, O, B, A, F, G, K, M, R, N, S. This curiously un-alphabetical order arises from errors by early spectroscopists, who confused some of the classes and also invented others that no longer exist. The standard mnemonic for remembering the order is Wow! Oh, Be A Fine Girl, Kiss Me Right Now, Sweetheart (although some astronomers prefer Smack! for the last letter).[1]

These classes diffuse into each other rather like the colours of a spectrum, and so they can be described only in generalities. W and O stars are the hottest, with temperatures of up to 100,000° C – about 15 times as hot as the Sun – which means that very few atoms can remain in their normal state. What is more, unlike the other spectral classes W stars give an almost entirely emission (bright line) spectrum. The Sun, as we have seen, gives an absorption spectrum (dark lines against a bright background), which means that the elements responsible exist in its atmosphere, and 'remove' their lines from the light emitted by the photosphere. This is true of most other stars as well. But evidently the atmospheres of the W stars, and, to a

[1] This brings to mind the equally irresistible mnemonic for remembering the order of the major planets: Many Volcanoes Erupt Mulberry Jam Sandwiches Under Normal Pressure.

lesser extent, the O stars, are themselves luminous because of the high temperature, and therefore give an emission spectrum. These two types are known as Wolf–Rayet stars, after the two astronomers who investigated them. Most Wolf–Rayet stars are very distant, so that despite their tremendous luminosity they appear rather faint and inconspicuous.

B stars (25,000° C) show characteristic hydrogen and helium lines; Spica, the blue-white brilliant in Virgo, is a typical example. A stars (11,000° C) are much cooler, with prominent hydrogen lines, and because of their lower temperature they are pure white; as we continue down the scale the colours redden, like a white-hot poker cooling, and for basically the same reason. A stars are also interesting for showing lines due to metals, especially magnesium, iron, titanium, and calcium, which strengthen in classes F (7,500° C; pale yellow) and G (6,000° C; yellow). Class G is noteworthy for including the Sun, and the solar spectrum is a mass of metallic lines.

In classes K (4,200° C; orange) and M (3,000° C; red), the temperature is so relatively low that atoms are sufficiently sober to combine and form molecules of certain compounds. Some of these are also found in sunspots, since they are regions of the photosphere at about the surface temperature of a K star. In classes R, N, and S (all roughly 2,500° C; red) molecular bands strengthen, particularly those of carbon compounds.

Spectral classification is obviously an extremely complex subject, and each class is in fact subdivided into 10 divisions (A_0–A_9, etc., with A_9 next to F_0). It is not necessary to probe very deeply into these characteristics. The important feature to remember is that all these apparent differences are due not so much to variations of composition, although these do occur to some extent, as to temperature. If we could somehow cool a B star down to 3,000° C we should get an M spectrum. It is temperature which brings different elements into prominence and eventually, in the very coolest stars, allows chemical compounds to form.

If a piece of wire is heated in a flame it glows more brightly as it heats up, and the same is the case with the stars: Wolf–Rayet stars are very luminous, while A stars are considerably dimmer – or, to use the more accurate expression, have a lower absolute magnitude. But what of the far cooler M stars? Betelgeuse, in Orion, belongs to this class, yet it is as distant as its neighbour Rigel (an A star) and appears roughly the same brightness. Obviously it must be of about the same absolute magnitude. How are we to reconcile this with the dimness of its surface?

The answer finally emerged in 1913, when two astronomers published a now famous chart: the Hertzsprung–Russell or H–R Diagram. What they did was to plot the spectral sequence (or temperature) of the stars against their absolute magnitude.

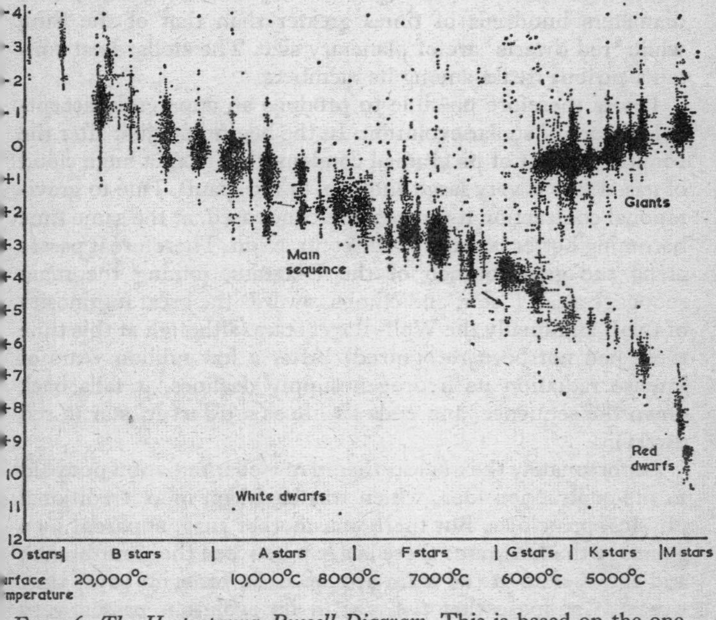

FIG. 36. *The Hertzsprung–Russell Diagram.* This is based on the one drawn by Gyllenberg, of Sweden; since each dot represents a star, its precise form clearly depends to some extent on the sample chosen. Notice how the giant stars are completely divorced from the main sequence; this division into two distinct classes is one of the greatest mysteries confronting astrophysicists. The Sun is arrowed.

A representation of the H–R Diagram is shown in Fig. 36. In addition to the 'main sequence', which we should expect, there were two classes of erratic stars which refused to conform to the behaviour of the majority: a few stars, of class M, were much too faint, while a great block of G, K, and M stars were many times more luminous than they should have been. Betelgeuse was far from alone in its defiance.

There could be only one answer. The super-luminous stars must be extremely large, to compensate for their feeble surface brightness, and the dim stars correspondingly small. These two classes, the 'giants' and the 'dwarfs', have now been fully confirmed. Some of the huge M stars, the 'red giants', have diameters hundreds of times greater than that of the Sun, while 'red dwarfs' are of planetary size. The stellar host contains curious freaks among its members.

It was therefore possible to produce an apparently acceptable theory of stellar evolution. In the beginning, just after the commencement of its gradual condensation from a huge cloud of gas, a star is very large and cool (a red giant). Due to gravitational contraction it grows more condensed, at the same time becoming hotter because of the contraction. Therefore it passes along the giant branch of the Diagram, joining the main sequence as an F star, and climbs towards the great luminosity of the B and finally the Wolf–Rayet stars (although at this time these had not been recognized). After a few million years of intense radiation its hydrogen supply declines, it falls back down the sequence, and ends its life as a dwarf M star (a red dwarf).

Unfortunately the nuclear theory of stellar radiation put paid to the contraction idea, which was based on more traditional physical principles. But there was another snag, apparent by a glance at the Diagram: there is a gap between the giant branch and the F stars of the main sequence. In order to cover this, we must suppose that the star in its evolution passed very rapidly through this adolescent stage, therefore explaining why so few stars are observed in this position. At one time, it was supposed that giants and main-sequence stars are completely unrelated. This idea is no longer held. Instead, it is supposed that giant and dwarf stars simply represent different stages of evolution. Astronomers today agree that the history of a typical star goes something like this:

Out of the interstellar gas and dust which exists in enormous clouds throughout the Galactic arms, at a rough density of about one hydrogen atom per c.c. and much lower values for helium and other atoms, condensations begin to occur. This condensation may, in the first place, be caused by radiation pressure from nearby stars acting on the 'dust' grains and giving them an initial velocity towards a common centre. Once the material has acquired sufficient adhesion, condensation proceeds due to gravity rather than to light-pressure, and collapse is more rapid. In anything between 100,000 and 10 million years, depending on the size of the initial cloud, the material has collapsed gravitationally to about a hundred-millionth of its size. At this point, the gravitational collapse is releasing so much energy that the protostar begins to glow. During its initial condensation, only heat waves are released. Later on, with increased energy available, the wavelength shortens to infra-red and finally visible light can be received by the astronomer's telescope. It is possible that this final stage in the evolution of the infant star is extremely rapid. Photographs taken of the same region in the Orion Nebula – undoubtedly a fertile birth-place for stars – have revealed several apparently new stars appearing in the space of a few years.

Our protostar has now become a star proper, and it takes its place on the main sequence according to its surface temperature. This temperature depends on the mass of the star, and it can be stated as a general rule that the more massive the star, the higher the temperature and the bluer its tint. A star of one solar mass will, on contracting to luminosity, join the main sequence where the Sun now is, in the G class. If the star has two solar masses, it will join its companions further up the main sequence, as an A star. Stars of three or four solar masses, which are the largest commonly found in the Galaxy, begin their adolescence as O and B stars.

The tendency of a star, once it has reached the main

sequence, is to remain near its original position while it begins converting its hydrogen into helium. As it burns its nuclear fuel it gradually brightens, but not very considerably, due to a slight expansion in size and therefore in surface area. The helium stock steadily increases, forming a hot central core, and the rate of growth of this core is critical in deciding how the star develops. The hotter the star, the more rapidly hydrogen is converted into helium and the faster the core grows. When the core reaches a certain critical size (with a mass of about 12% that of the entire star) it has to stop growing, because there is insufficient insulation left in the outer hydrogen layers to hold it at the temperature of about 20 million °C which allows the nuclear processes to continue. The time taken to reach this condition depends on the mass of the star. The Sun, about 4,500 million years old, is still on the main sequence with a relatively small core. An O or B star, on the other hand, may take only a few million years to reach the end of its initial hydrogen-burning life.

What happens now is that the helium core collapses inwards on itself. The pressure raises the central temperature, and the new release of energy throws the outer envelope of the star out in an expanding shell. As the size of the star's surface increases, so its outer temperature and therefore surface brightness fall, so that the star reddens in colour and becomes a giant. The cluster of red giants in the diagram has evolved principally from main-sequence stars of the F and G class, while the rare, luminous early-type stars evolve into blue or white giants, the searchlights of the Galaxy.

This is not the end of the story. It is probable that the more massive stars develop such a large, unstable core that it collapses inwards on a scale sufficient to rip off the outer skin of the star in the blast of fury that we see as a 'nova', the remaining core cooling and collapsing into the white dwarf state, which seems to be almost the final stage in the history of all stars. Quieter stars, like the Sun, will go through their entire development on a more peaceful scale, but the end result will be the same. White dwarfs are planet-size bodies containing almost all the original star's mass, and hence are extremely dense; but it seems like that the recently-discovered pulsars may represent the ultimate in dwarfs: 'neutron' stars, in which the star's material is contained in a sphere perhaps less than 10 miles across and spinning once a second!

The question of stellar masses is an interesting one, and must obviously have some bearing on evolutionary theory. Star luminosities vary tremendously; the brightest known star is over a million times brighter than the Sun, though it is so far away that it appears very faint, while one of the dimmest, known as Wolf 359, is only 1/50,000 as bright – it is, as might be expected, a nearby red dwarf. But masses are nothing like so diverse, ranging from 100 times that of the Sun to about $\frac{1}{25}$. Clearly, then, a huge star like Betelgeuse, with a diameter of 200,000,000 miles, must be in a very rarefied state, the density of its outer reaches being like that of a comet's tail. Conversely a white dwarf such as Kuiper's Star, with a diameter of 4,000 miles, is incredibly compact; a cubic inch of its material, if brought to the surface of the Earth, would weigh 1,000 tons! It is literally atomic, the atomic nuclei being so tightly packed together that there is no space between them. This remarkable material is often referred to, rather inaccurately, as 'nuclear fluid'.

Thus far we have talked glibly of stellar luminosities without explaining how they are worked out. A star's apparent magnitude can be calculated easily enough, but if we wish to find its absolute magnitude, using this as a key, its distance must be known. The problem of gauging the distances of the stars is a fascinating one.

When Eros swung near in 1931 observatories all over the world took photographs to measure its precise position against the stellar background. Then, knowing the distances between the various stations, the parallactic shift of the tiny planet could be used to work out its distance from the Earth. This method, however, was possible only because of its unusual closeness. Even the nearer major planets reveal almost negligible shifts, and so obviously the largest terrestrial baseline is utterly useless when it comes to dealing with even the nearest stars. What is needed is a much longer baseline. Short of soaring off the Earth altogether, the only way is to use the Earth's motion in space. In other words, if we observe the position of a star we imagine to be relatively close to us against the background of stars that are at infinity (or so we hope), and then wait six months until the Earth is at the opposite extreme of its orbit and re-observe the star, we shall have brought into play a baseline equal to the diameter of the Earth's orbit: 186,000,000 miles (Fig. 37). The resultant shift is known as a trigonometrical parallax, and this was the earliest method used.

The first star to be honoured by having its distance measured was a naked-eye object in Cygnus, No. 61 in Flamsteed's

FIG. 37. *Trigonometrical parallax.* Needless to say, the angle is vastly exaggerated. If the star in question were α Centauri, the nearest to the Sun, its distance from the Sun in the diagram would have to be nearly ¾ of a mile! It is a tribute to astronomers that such tiny angles – and some far smaller – can be measured with considerable accuracy.

catalogue and therefore known as 61 Cygni. The reason for its choice is an interesting and important one. For, contrary to popular belief, the stars do not remain absolutely still. In fact they are all flying around in space at velocities of many miles per second, but their distances are so colossal that these drifts are hardly noticeable; the result is that the night sky of 1,000 years ago would seem, to the casual eye, to duplicate its present aspect. But minute observation has proved that every star has a certain 'proper motion', or slight movement relative to its fellows. Clearly, the nearer stars are likely to show the greater movement, and 61 Cygni, moving across the sky at a rate sufficient to cover the Moon's apparent diameter (½°) in three centuries, became known as the 'flying star' when its motion was brought to light by Piazzi in 1792. Here was a splendid candidate for distance measurement.

The first assault on the 'flying star' was made in 1837 by a brilliant German astronomer, F. W. Bessel. Using a new type of instrument that allowed extremely accurate measurements of position to be made, he assiduously observed 61 Cygni for a year. In November 1838, a momentous date for stellar astronomy, he announced that the star showed a trigonometrical parallax of $\frac{1}{3}''$.[1] Later observations increased this to nearly $\frac{1}{2}''$. A simple calculation inferred its distance as 67,000,000,000,000 miles, or 11 light-years. For the first time a true yardstick had been plunged into space.

Now $\frac{1}{2}''$ is a minute angle, equivalent to the diameter of a new penny seen from a distance of 7 miles! Two months after Bessel's announcement, Henderson at the Cape of Good Hope published a parallax of almost 1″ for the bright star α Centauri, equivalent to a distance of $4\frac{1}{3}$ light-years, and this still tiny shift has proved to be the largest stellar parallax; the system of α Centauri, which actually consists of three stars revolving around each other, is the Sun's nearest companion in space. To return to the scale suggested in Chapter 1, they may be compared with two oranges separated by 1,400 miles, while even 61 Cygni, another extremely close neighbour, is 3,600 miles away.[2]

It is a tribute to astronomers that such infinitesimal evidence can be turned to so impressive account; a sobering thought that the microscope plumbs the depths of space, since all modern measures are made from photographic plates. But clearly, the use of trigonometrical parallaxes is very limited. At 50 light-years the shift has become unbelievably minute; at 500 light-years, imperceptible. Clearly, other methods must be sought; for the 6,000-odd stars whose distances have been gauged by this method, while an impressive achievement, are a poor proportion of the millions that throng the Galaxy.

[1] A degree (°) is divided into 60 minutes (′), which in turn are divided into 60 seconds (″). 1″ is therefore equal to 1/3,600°, or about 1/1,800 of the apparent diameter of the Moon.

[2] Determination of distances by the parallax method has given rise to a new unit, the parsec. One parsec is the distance at which an object would show an annual parallax of exactly 1″ (rather less than the distance of α Centauri), and is equivalent to about $3\frac{1}{4}$ light-years. However the latter unit is more familiar, and is used throughout this book.

There is another drawback also. When measuring the six-monthly displacement of the star against the stellar background, an assumption must be made about the comparison stars from which the displacement is measured. If they are at infinity, the measured parallax will be a true one. But suppose they too are showing a slight parallax? In this case the measured shift of the star will be slightly too small, and while the effect is likely to be negligible in the case of near stars, it adds an extra uncertainty to parallaxes obtained near the limit of measurement. So how else can we tackle the problem?

Suppose we see an inaccessible street lamp a considerable distance away and want to find out just how remote it is. Luckily there is an identical lamp nearby, whose distance we can measure easily. It comes to exactly 10 yards.

We then take a photometer, which is an instrument for measuring the brightness of a luminous object, and compare the 'apparent magnitudes' of the two lamps. The ratio works out to 100; in other words, the nearer lamp appears 100 times as bright as the more distant one. But in fact we know that their luminosities or 'absolute magnitudes' are the same. Since brightness falls off with the square of the distance, the farther lamp must be $\sqrt{100} = 10$ times as far away. $10 \times 10 = 100$ yards.

It is clearly not necessary for the two lamps, or the two stars, to be of the same brightness; as long as we know the absolute magnitude of the distant star, we can calculate its distance by comparing it with another star (real or imaginary) whose absolute magnitude and distance are known. But obviously there must be some means of accurately scaling absolute magnitude. We speak of lamps in terms of candlepower, but a candle does not put up much of a show on the stellar scale! What is done is to imagine the stars as seen from a standard distance of 32·6 light-years (10 parsecs), and to define their absolute magnitude as their apparent magnitude when seen at this distance. Spectacular things happen when we start to rearrange the stars in this regimentation. 61 Cygni disappears from the naked-eye range altogether; Betelgeuse flashes out as bright as Jupiter, and another red giant, Antares, rivals Venus! But most startling of all is to find that the Sun is nothing more than a dim yellowish speck. Its absolute magnitude is 4·7, so that at the standard distance it is definitely one of the less dramatic stellar members.

This is all very well, but there is still the problem of finding the star's absolute magnitude. The clue lies in its spectrum. We have already seen, from the H–R diagram, how main-sequence stars decrease in luminosity as we pass along the scale. A B star, for instance, will have an absolute magnitude round about -5, while a red dwarf is nearer $+9$. These are only rough estimates, but detailed analysis can usually reduce the value to within quite fine limits. Once we have found the absolute magnitude of the star, we are back on the ground of the street lamp problem. We know how bright the star would appear at a distance of 32·6 light-years, and we know how bright it actually appears. The rest is simple calculation.

These 'spectroscopic parallaxes' offer fresh hope to the astronomer. There is almost no limit to the faintness of a star whose spectrum can be imprinted on a photographic plate, and once it is secured the class and absolute magnitude can be inferred. However, we must not be over-optimistic. Many stars have spectra that do not readily yield to accurate measurement, especially the very luminous stars at the upper end of the sequence, and this once again limits our probes, since at very great distances only the most luminous stars can be made out. Moreover there is another snag, for space itself is not completely empty. Throughout the Galaxy is spread an inconceivably tenuous diffusion of dust, and when the distances are sufficiently large this dust has an appreciable effect on the starlight, reddening and dimming it at the same time. As a result the apparent magnitude of the star is fainter than it would be were space perfectly transparent, and this upsets the measures. Once again 'laws of reddening' have been invoked, but matters are complicated by the fact that absorption is not equally intense in all directions.

Despite these complicating factors, spectroscopic parallaxes are now the main method by which astronomers gauge the distances of individual stars. There are other methods as well;

the Sun is moving through space at 11 miles per second, carrying the Earth and the other planets with it, and this can be used as an ever-lengthening baseline from which to measure parallaxes. However, this has to take into account the fact that all the stars are travelling at considerable velocities, which is in itself a matter of great interest.

Let us take a simple situation. We are out at sea, and a ship is sailing between us and the distant coastline. By watching its motion relative to a landmark its drift at right-angles to our line of sight can be estimated, and by noting its increase or diminution of size its velocity towards or away from us can also be guessed. By combining these two speeds its real motion can be worked out.

The position of an astronomer trying to find a star's true

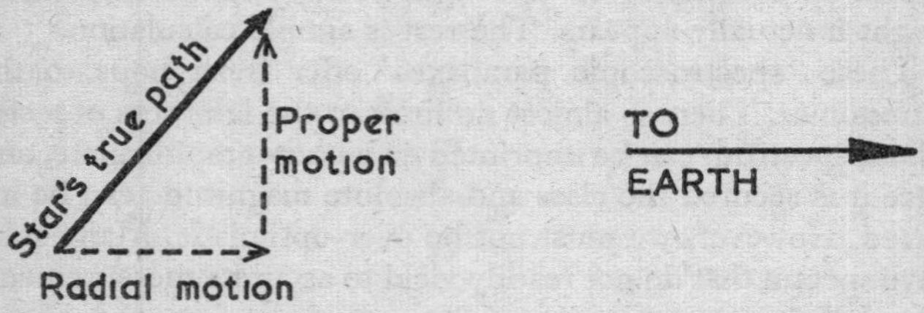

FIG. 38. *Three sorts of motion.* Astronomers have to calculate a star's true path in terms of its radial and proper motion.

path of motion is rather analogous to that of the seafarer. We must remember that in all cases we see stellar (and planetary) motions not in their three-dimensional aspect, but projected against the sky. In the case of 61 Cygni its annual drift amounts to just over 5″, and once its distance is known its speed *across our line of sight* can be calculated. But it does not follow that it is actually pursuing this path. In fact it is very likely, indeed almost certain, to be also advancing or receding (Fig. 38). In the case of the ship this so-called 'radial motion' could be judged by its change of size. But the stars are too distant to show physical disks at all. Direct observation cannot, therefore, establish a star's radial motion; all a telescope can do is measure its proper motion across the stellar background.

Once again resort must be made to that priceless instrument the spectroscope, and the phenomenon it can reveal; the Doppler Effect. This has already been mentioned in the case of Venus, where a sufficiently rapid rotation would produce a shift of the spectral lines towards the blue (if the limb were approaching), or the red (if it were receding). The leisurely Venus refuses to betray any shift at all, but stars are much more active; they are all travelling through space at velocities of several miles per second. If any or all of this motion is directed towards or away from the Sun, the Doppler shift will reveal it.

What is more, the shift immediately gives the radial velocity. In the case of proper motions it is also necessary to know the distance of the star, but Doppler shift measurements are independent of distance. In this way the radial velocities of distant galaxies can be determined even though these galaxies are too remote for very precise estimates of distance. This is an absorbing subject that must be reserved for a later chapter.

The results of work done on stellar motions more properly belong to Chapter 23, since it is bound up with the nature of the Galaxy as a whole; it is sufficient here to mention the technique involved. However, there is a universal proper motion spread over the entire sky which is not really a galactic phenomenon at all: it is due to the motion of the Sun, and it was first established by the industrious Herschel in 1783.

If we are walking through a forest, the trees in front of us appear to spread out as we approach, while those behind close in – this is simply an effect of perspective. Herschel reasoned that since many stars show proper motion, there was no reason why the Sun itself should not also be travelling through space. In this case the stars towards which it was travelling would move apart, there being a compensating congregation in the opposite direction. A careful study of stellar proper motions, spread over the sky, should reveal this drift if it were there; and his hopes were splendidly fulfilled. He found the direction of the Sun's motion, termed the 'solar apex', to lie in Hercules, and subsequent work has shown his position to be amazingly near the truth. The Sun is in fact moving towards the region of the brilliant star Vega.

Needless to say, there is not the slightest danger of a collision. For one thing, Vega is 26½ light-years away, which makes the

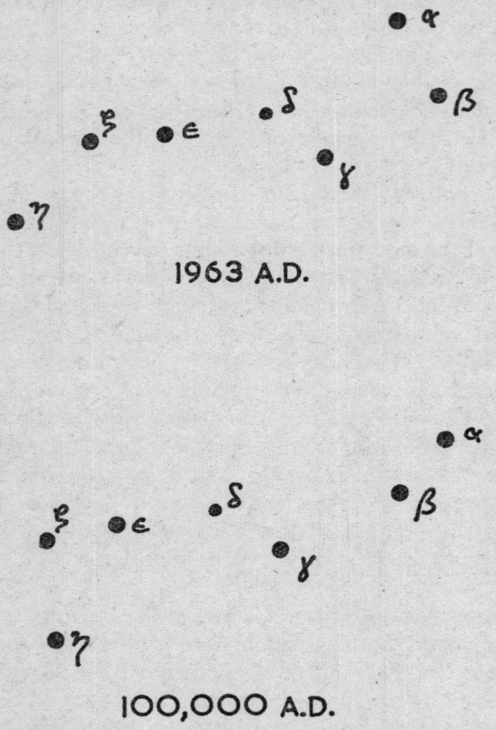

FIG. 39. *The Great Bear in AD 100,000.* All the stars except α and η are moving as a group.

Sun's velocity of a mere 400,000,000 miles per year a hopeless crawl; it will take hundreds of thousands of years to cover the distance, and in any case Vega's own proper motion will carry it well out of the way.

Proper motions are gradually changing the face of the

heavens, though so slowly that thousands of years must elapse before the general effect becomes noticeable. The most obvious effect concerns the brighter constellations, and Ursa Major is a case in point. We have said that many stellar motions seem to be random, but some stars do in fact belong to groups. Five of the seven bright stars in the Great Bear are moving in more or less the same direction, while the two extreme members, α and η, are going the opposite way, and in 100,000 years the constellation will present a markedly different aspect.

In this chapter it has been impossible to more than touch on a few of the many important departments of stellar astronomy. How were stars born? Why are they divided into different classes? What happens to them when they die? There are many theories, but few universal conclusions. In fact the chief certainty of our knowledge of the stars themselves is its own uncertainty.

CHAPTER 19

Double Stars

IF WE take a large sheet of paper and draw a dot in the centre to represent the Earth, subsequently scattering grains of sand over the rest of the paper to represent stars, it is to be expected, on the laws of chance, that some of these grains will fall so that they appear almost directly in line with the 'Earth' (Fig. 40). The real stars are distributed in space in a similarly random way, with the obvious difference that theirs is a three-dimensional scattering. So once again we should expect instances

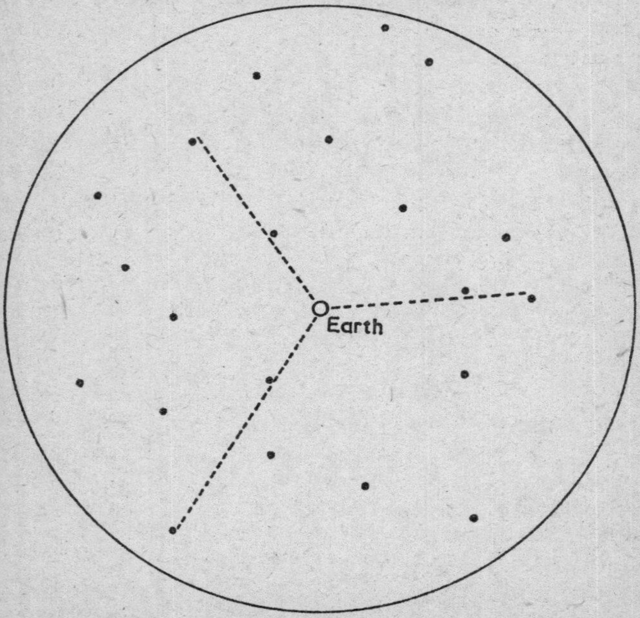

FIG. 40. *Optical doubles.*

where two or even more stars appear side by side in the sky, whereas one is in reality far more remote than the other, and in no way connected with it.

The most famous double star in the sky is of this type: closely north-east of the star ζ Ursae Majoris (Mizar) is a much fainter star (Alcor), which can be seen easily with the naked eye. A glance at the bright star α in the constellation of Capricornus shows that it consists of two roughly equal stars. These are two examples of naked-eye double stars, and there is no mystery about them; in both cases they are simply line-of-sight effects. These are called 'optical' doubles.

But double stars persist when a telescope is turned to the sky. In fact they not only persist, but become more and more numerous. About a tenth of all the stars in the sky, when examined closely, have faint companions within a few seconds of arc, and some are so close as to be visible only with the largest telescopes. In short, there are far more than could possibly be expected on the grounds of luck alone.

Herschel, who made double stars his special province, soon realized this; during his reviews of the heavens he discovered hundreds. In 1802 he wrote that 'casual situations will not account for the multiplied phenomena of double stars', and in the following year he published evidence that some doubles consisted of stars revolving around each other. The science of true doubles, or 'binary' stars, was born. Not only that, but their slow waltzes in the sky were beautiful proof of Newtonian theory applied to the universe as against just the solar system.

To the casual amateur searching for pretty sights, it does not matter whether the double is optical or a binary system; to professional astronomers, however, optical doubles are of little interest. All that they can be used for is checking the proper motions of the components, whereas observation of a binary system and determination of the orbit leads to accurate knowledge of the masses of the stars and other information of tremendous value. There is no room to go into the techniques of this type of work, but we can at least mention some of the many interesting binaries that have come to light over the years.

Amateur observers have compiled lists of the most spectacular doubles for small telescopes, and most of these are attractive because of their contrasting colours; star tints vary widely, and when two different hues occur in a double the contrast affords a pleasing spectacle. By common consent the most beautiful double star in the sky is β Cygni (Albireo), at the foot of the 'cross'. To the naked eye it appears an ordinary star of the 3rd magnitude, but a small telescope converts it into a rich yellow star with a fainter bluish companion

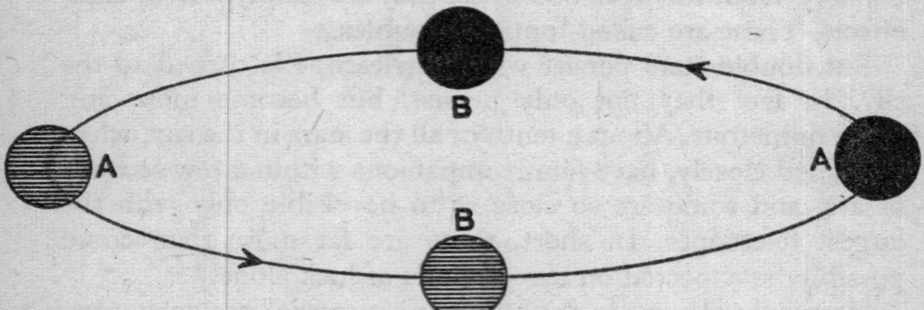

FIG. 41. *Inclined view of a binary system.* Even if the two members of a binary system are revolving around each other in a circular orbit, the chances are that we see its plane at a more or less inclined angle. They therefore seem far apart at A and close together at B.

at a distance of 35″. It is a genuine binary system, but the movement is inconceivably slow; the period is to be measured in thousands of years. It prompts the question of why two such dissimilar stars should have been formed together. The primary, as we should expect from its colour, is a K star, while the companion, which falls into the B class, is very much hotter.

Albireo remains virtually the same from century to century, but there are other bright doubles which form binaries of much shorter period. Perhaps the most famous is Castor (α Geminorum), with a period of about 350 years. Its components were at their widest separation in 1880, when their distance was $6\frac{1}{2}''$; since that time they have been moving closer together, and the present distance is less than 2″.

This does not mean that the components are actually approaching each other, although this does happen to some extent, since their orbit is not circular. The main reason is that we are seeing the orbit at a very inclined position, so that the stars seem to swing together and apart (Fig. 41). Two other bright Castor-type stars are γ Leonis (400 years) and γ Virginis (180 years). These are both easy telescopic objects, and it is significant that in all cases the components have almost exactly the same spectral class. This is what we should expect if, as seems likely, they were formed together under the same conditions.

In a way it is misleading to speak of a binary's orbit as a straightforward circle or ellipse, since this suggests that one star revolves around the other. In saying this of the planets' motion round the Sun, or the Moon's round the Earth, it is more or less true, because in each case the main member of the system has tremendous superiority of mass. For instance, the Moon and the Earth revolve around their common centre of gravity, but the Earth is so much the more massive of the two that this point is beneath its surface; it therefore wobbles like a cam while the Moon sweeps out a large orbit. Things are very different in the case of binary systems. As we have seen, stars all have the same order of mass, and in general the centre of gravity will be roughly midway between the components. So we can no longer conveniently speak of one revolving around the other; the two move together rather like the masses on a dumb-bell when it is twirled in the hand.

Many bright stars have faint companions. A spectacular example is the red giant Antares (α Scorpii) which, with a diameter of 480,000,000 miles, is one of the largest stars known – it is in fact termed a 'supergiant'. At the same time its mass is only 30 times that of the Sun, so that it is exceedingly tenuous. The companion is blue, but contrast with the bright star makes it seem green, with spectacular results. We see the same sort of thing in another star, α Herculis.

These doubles, and hundreds of others, were discovered many years ago – mainly by Herschel, who dredged the northern sky of its brighter twins. Following in his footsteps, other workers have taken up the pursuit, and the total number of doubles known today cannot be much less than 30,000, the vast majority being faint, close binary systems. Most of these were found simply by patient observation, and it is therefore well worth telling the story of the companion that, like Neptune, was known to exist before it was seen. It belongs to Sirius, the brightest star in the sky, and the investigation was made by the observer of the 'flying star', Bessel.

After observing the very large proper motion of 61 Cygni, Bessel turned his attention to Sirius, which also showed a considerable drift, although not so large. But in 1834 he noticed a very curious fact. The motion of Sirius was not regular; it slowed down and accelerated again in a rather drunken way, and after ten years of careful work he announced that Sirius must have an almost equally massive companion, the two forming a binary system with a period of half a century. He formed the same conclusion with regard to the nearby star Procyon (α Canis Minoris), which is fainter but still one of the brighter stars.

Bessel had proved their existence, but no observer had yet succeeded in spotting either of these mysterious companions. We must remember that nothing was then known of the enormous range of stellar brilliance, and since the masses of these companions were apparently comparable with those of their primaries, it was unthinkable that they should be too dim to see. Bessel died in 1846, his 'invisible stars' unconfirmed, and six months before the triumphant vindication of another branch of mathematical astronomy in the discovery of Neptune. He had in fact been working on this problem, and it would have been interesting to know his reaction to this feat.

Sixteen years later two of the most famous of all telescope makers, the Clarks (father and son) were testing a new large telescope of 18½ inches aperture, then one of the biggest in existence. Alvan, the son, decided to test it on Sirius. It so happened that the bright star was obscured by a wall, and upon emerging into the view of the telescope it was seen to be preceded by a faint companion. Sirius B,[1] had been accidentally

[1] In the case of binary stars the bright and faint members are termed A and B respectively.

discovered in precisely the spot indicated by Bessel. The Clarks could not have wished for better evidence of the excellence of their new telescope. The date was January 31st, 1862 – just over a century ago.

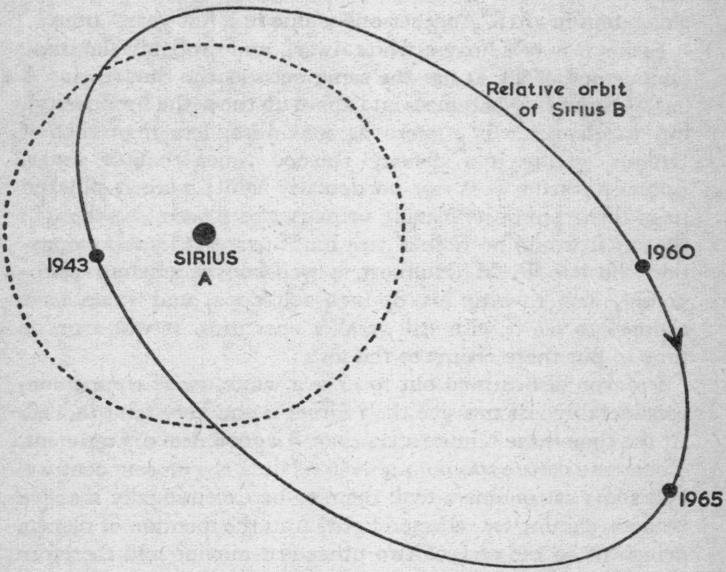

FIG. 42. *Sirius A and B.* Although the two stars are revolving around their common centre of gravity, it is in most cases more convenient to consider the brighter member fixed in space; this is because an astronomer always measures the position of the fainter star relative to the primary. The result is the 'relative orbit'. The dotted line represents, very approximately, the limits of the glare around Sirius when viewed through a large telescope, which occurs mainly through atmospheric disturbances. For this reason the companion is lost from view around the time of closest approach.

Fig. 42 shows the relative orbit of Sirius B around its primary; we say 'relative' because although the stars are moving about their mutual centre of gravity, it is far more convenient to consider Sirius A fixed. The period is within a few weeks of 50 years, which makes Bessel's prediction still more remarkable. For much of this time the companion is very close to Sirius A and lost in the glare, and at the time of Bessel's announcement it was hopelessly near its primary. By 1862 it had swung away almost to its maximum distance of 11″, but by 1890 it was lost once again, reaching its next elongation in 1918. Another one is due in a few years' time.

Sirius B is of course a white dwarf, and probably the most famous one of all. It has the same mass as the Sun (Sirius A has 2½ times the solar mass and about 26 times the luminosity), but its diameter is a mere 24,000 miles, less than that of Uranus, giving it a density 100,000 times that of water. Actually, Sirius B is not particularly faint; were it plucked from the grasp of its blazing primary and placed elsewhere in the sky it would be visible with binoculars. At its last elongation the late F. M. Holborn, a well-known amateur astronomer, saw it with his 8½-inch telescope, and some have claimed to see it with still smaller apertures. It will soon be time to put these claims to the test.

Procyon also turned out to have a white dwarf companion, considerably less massive than Sirius B and very faint indeed. At the time these minute stars caused a good deal of argument. Their true nature was not understood until the present century, and some astronomers took them to be exceptionally massive planets, shining by reflected light! And the mention of planets brings us to the case of two other fast-moving and therefore nearby stars, 70 Ophiuchi and our old friend 61 Cygni.

61 Cygni is the more interesting of the two. It is, in fact, an easy double star – a genuine binary system, with a period of about 700 years. A Swedish astronomer, Dr K. A. Strand, announced in 1944 that he had discovered the fainter component to show a distinct 'rippling' motion in addition to its orbital movement, thereby confirming suspicions held 50 years earlier. The case was much the same as with Sirius A, except that the companion to 61 Cygni B was clearly a lightweight. In fact Strand placed its mass as only 15 times that of Jupiter, or $\frac{1}{70}$ of that of the Sun. This has far-reaching implications. What we know of stellar formation tells us that so insubstantial a star must be an extremely rare phenomenon, and because of

its invisibility the logical conclusion is to infer a non-luminous body: a planet. By the same chain of reasoning the planet revolving around one of the components of 70 Ophiuchi, another binary star, has a mass equivalent to that of only 12 Jupiters, while Barnard's Star, which is only 6 light-years away, is thought to possess two bodies of planetary mass.

Large as they are, we can never hope to see these planets from the Earth; not only are they enormously remote, but the suns around which they revolve are much less luminous than our own. They must therefore be cool, twilight worlds, probably 'giants' in composition; perhaps even the chief members of planetary families resembling the solar system.

We have already said that there is no reason why solar systems should not be common phenomena, and these revelations of 61 Cygni and its fellows give food for thought. Planetary perturbations must always be very small, and because of this there is no hope of detecting them on a remote star; both these stars are near neighbours of the Sun, their distances being 11 and 16½ light-years respectively. If we draw a sphere around the Sun with a radius equal to the latter distance we shall trap only 53 stars in all, and of these three are known to be planet-bearing. If this proportion is a correct average (or even if it be 10 or 100 times too great), it still leaves room for a million planetary systems in the Galaxy. Among this host duplicates of the Earth cannot be rare.[1]

The realm of interstellar planets must surely belong forever to Bessel's 'astronomy of the invisible', but double stars themselves give birth to another branch of this same mysterious science. A great many stars, among them the two components of Castor, are known to be double and yet are too close to be separated into their components with any telescope. Altogether over a thousand of these very close binaries are known; they

[1] In April 1963 Dr van de Kamp, director of the Sproul Observatory, announced the discovery of a third extra-solar planetary system. From a study of photographs taken of Barnard's Star, which is a mere 6 light-years away, he inferred the existence of a planet 1½ times the mass of Jupiter, revolving at a distance of about 400,000,000 miles. Clearly, this latest discovery raises enormously the already strong likelihood of there being many millions of planetary systems in the Galaxy.

are called 'spectroscopic' binaries because it takes a spectroscope to discover them.

Suppose the two components are revolving as shown in Fig. 43 (for convenience the orbit is taken as circular), and we, on the Earth, see the system more or less edge-on. In position 1 star A is approaching while star B is receding, and in position 3 the reverse is the case. In the intermediate state, 2, the stars are moving more or less across the line of sight and show no radial motion at all.

Let us assume, for convenience, that the stars are in all ways

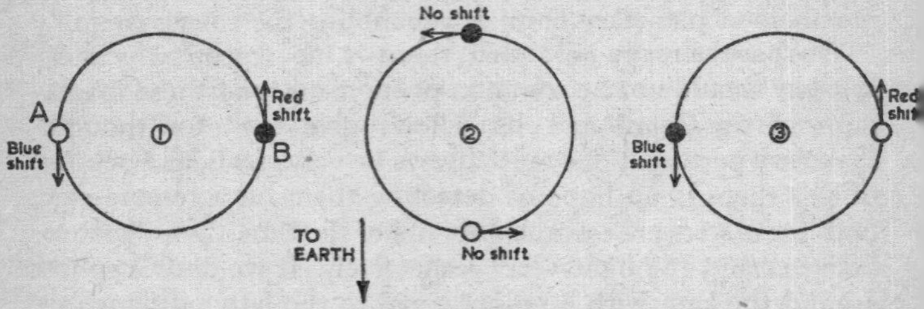

FIG. 43. *A spectroscopic binary.*

identical and produce the same spectrum. Because they are so close, a spectroscope shows the combined spectrum of both components. But we must remember that for most of the time they have opposite radial velocities, and these produce a shift of the spectral lines. For example, in position 1 star A is showing a blue-shift, star B a red-shift, and as they run through their period the opposite shifts of the components will give a spectrum in which the lines widen, eventually divide, and then close again. It is obvious that in position 2, where there is no net radial motion, the two series of lines will coincide exactly.

The first spectroscopic binary to be discovered was β Aurigae, in 1889, and the number is steadily increasing. Since the components, to be inseparable telescopically, must be relatively close, the periods are usually short; most are less than 100 days. The components of Castor have periods of about

3 and 9 days, while a third star, detectable with a small telescope, also forms part of the system and is itself a close binary! Castor therefore consists of six stars gravitationally connected. There are other examples of these amazing multiple systems, but this masterpiece of complication forms a fitting climax to our survey of double stars.

CHAPTER 20

Variable Stars

IN THE winter constellation of Perseus there is a star which winks. For 2½ days at a stretch it remains at almost constant brightness, but in the next 5 hours it dims by over a magnitude. Another 5 hours sees it returning to its original lustre, followed by a further 2½ days of relative stagnation. It is a tribute to the old Arab astronomers that they noticed it and called it Algol. Algol means the Demon Star, and their reasons for applying the name to β Persei are clear enough.[1]

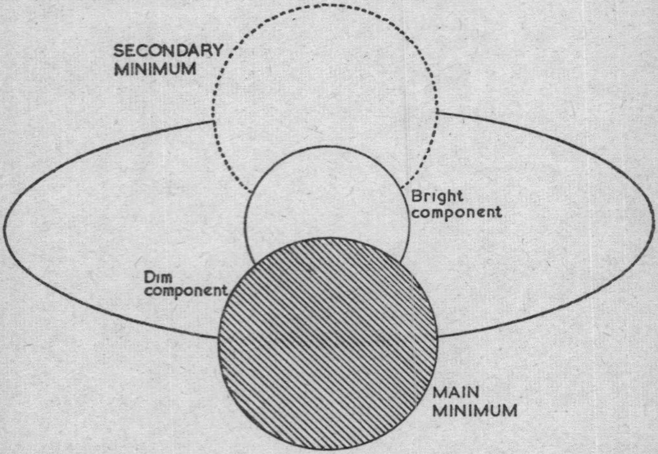

FIG. 44. *Why Algol winks.* This diagram shows the relative orbit of the fainter component around the brighter.

There are many demon stars, or variable stars, in the sky, but not all are as regular as Algol. This is because there are two basic reasons why a star appears to vary in brightness.

[1] Oddly enough, some scholars claim the association to be purely fortuitous, and that the Arabs were unaware of Algol's caprices.

Algol, in fact, is a cheat, for it is a spectroscopic binary whose orbit we see nearly edge-on. One of its components is much larger and dimmer than the other, so that when it cuts across

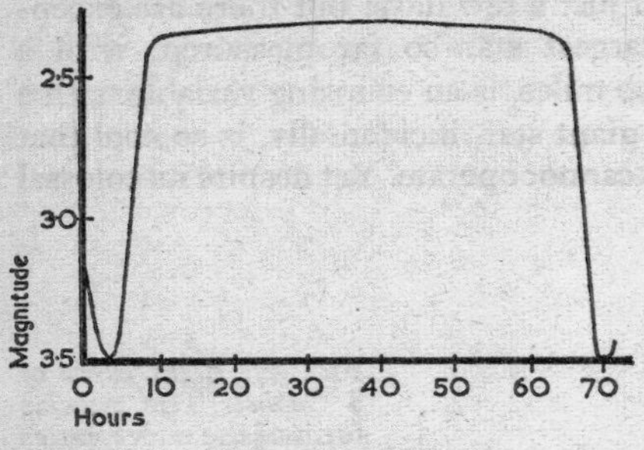

FIG. 45. *Algol's light curve.*

and eclipses its fellow, there is a marked diminution of light (Fig. 44). However, there must also come the time when the bright component passes across its companion. In this case it cannot cover it so completely, and so there is only a slight fall in magnitude: a 'secondary minimum'. The light curve is shown in Fig. 45. Algol belongs to the class of stars known as dark-eclipsing variables, and they obviously must reproduce these phases in an interminable rhythm.

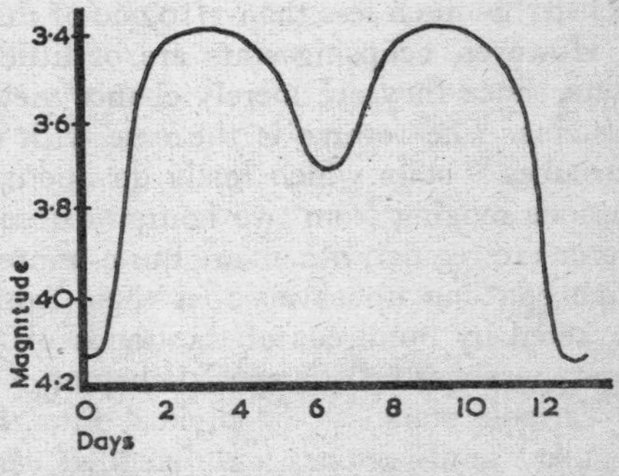

FIG. 46. *Light curve of β Lyrae.*

There is another class of eclipsing binary, typified by the bright β Lyrae. Here the components are about equally bright, so that the secondary minimum is more conspicuous and we get two falls for the price of one (Fig. 46). The bridge between bright-eclipsing and dark-eclipsing variables is clearly a very diffuse one. Most of them, in keeping with other spectroscopic binaries, have periods of just a few days, but there are exceptions; ε Aurigae, the largest star so far measured, with a diameter of 1,800,000,000 miles, is an eclipsing variable with a period of 27 years. The giant star, incidentally, is so cool that normal nuclear reactions cannot operate. Yet despite its colossal

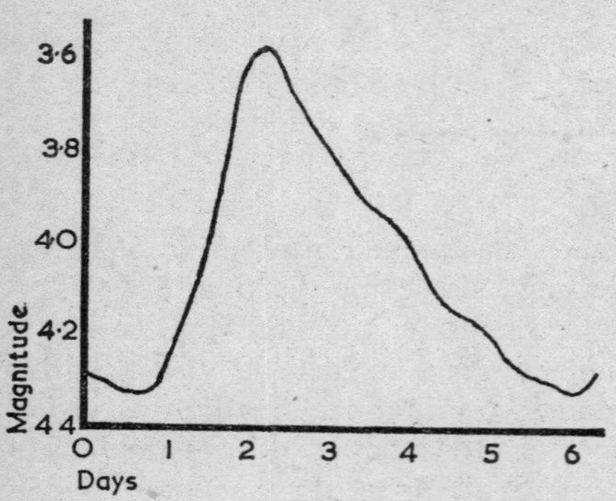

FIG. 47. *Light curve of δ Cephei.* The precise form of the curve varies with different Cepheids, but in almost all the rise to maximum is more rapid than the subsequent fall.

size its mass is only 18 times that of the Sun, so that its outer density is much less than 1/10,000 of that of ordinary air.

However, eclipsing stars are of little interest as variable stars, since they are merely chance views of ordinary binary systems. The reverse is the case with some of the genuine variables – stars which really do fluctuate in brilliancy over periods ranging from two hours or even less, to many years. Some are regular, others are quite unpredictable; some suffer such startling upheavals that their light output is suddenly rocketed by hundreds of thousands of times, and these, the novae, are worthy of a special chapter on their own.

Variable stars can be divided into three distinct classes: regular, semi-regular, and irregular, and it is the regular variables which are of the greatest service to astronomers. In fact, it is to some of these that we owe much of our knowledge of the universe.

One of the most important discoveries of the century was the detection of a class of variable stars known as Cepheids. Their prototype is the naked-eye star δ Cephei, which varies from magnitude 3·6 to 4·3 and back in a period of 5 days 9 hours, and its regularity, like that of an eclipsing variable, is precise to a fraction of a second. Most Cepheids are too distant to be seen without a telescope, but about a dozen are visible with the naked eye, among them the Pole Star, whose range of magnitude is, however, too small to be at all noticeable.

The periods of Cepheid variables range from about 25 hours to 45 days, but most cluster about the 7-day mark – we might call them celestial commuters! What is so interesting and important is that their absolute magnitude is intimately related to the period; the so-called Period-Luminosity Law. In other words, if we time the period of a Cepheid we can find its absolute magnitude, and this, as Chapter 18 explained, enables the distance to be worked out. On average, the longer the period the brighter the star.

Cepheids thus constitute a fresh and powerful tool for the astronomer anxious to probe into space. Trigonometrical parallaxes are limited to the Sun's near companions, and spectroscopic parallaxes to stars leaving a suitable spectrum, as well as being above a certain apparent magnitude. But Cepheids are much more tolerant. Provided they are bright enough to be detected at all their absolute magnitude can be found and the distance calculated. It is also fortunate that they are mostly very luminous stars, some being more than 500 times as bright as the Sun, which means that they can be detected at great distances.

Cepheids do not cluster together in any special way; they are scattered throughout the Galaxy, and we have as yet no idea why they should all conform so faithfully to the Period-Luminosity Law. There are actually two distinct classes. One, the 'classical' or Type I Cepheids, inhabit the spiral arms, while a more recently-identified family, Type II Cepheids, are to be found in the nucleus. These obey the same broad relationship, but are rather fainter, and until the difference was realized they caused considerable confusion among distance estimates.

So much of astronomy is interconnected that it is impossible to keep its departments entirely divorced, and no mention of the Cepheids would be complete without reference to their extreme value not only to the Galaxy, but to the universe as a whole. It is therefore necessary to anticipate a part of Chapter 24 and mention how these extraordinary stars were and still are used to gauge the depths of intergalactic space.

The Galaxy, with its 100,000,000,000 stars, one of which is the Sun, is but one of literally millions of millions of other galaxies distributed throughout the volume of the observable universe. It was a long time before this was realized; even at the turn of the century it was thought that other galaxies, if they existed at all, were too far away to be visible even with a powerful telescope. But gradually astronomers realized that some of the nebulous clouds visible in the night sky, instead of being relatively nearby masses of glowing gas, are really immensely distant galaxies. One of the brightest is visible with the naked eye in the constellation of Andromeda, and so must be one of the closest. How far away was it?

Here was an irritating position, for at their great distance the individual stars were too faint for spectroscopic analysis of their brightness. The breakthrough came in 1925, when the renowned astronomer Edwin Hubble discovered Cepheid variables faithfully winking away among the massed swarms of stars; altogether he observed about fifty. They fitted the period-luminosity curve for Cepheids belonging to our own Galaxy, and were evidently of the same type; therefore their absolute magnitudes could be found. The distance of the Andromeda galaxy worked out at 800,000 light-years. It was the first step towards understanding the true vastness of the universe, and as other galaxies came to light so the faithful Cepheids revealed themselves and kindly provided the key to their distance.

The shock came in 1952, when the late Walter Baade began to have his suspicions about the Andromeda galaxy. Luminous stars of a type known in our own system refused to show up. There was no reason to suppose that they did not exist; the obvious alternative was that the galaxy must be considerably more remote than we had supposed, and this led to the discovery of the different classes of Cepheid. The formula used to calculate the distance belonged to what are now recognized as the less luminous Type II Cepheids, whereas the variables observed in the Andromeda galaxy were the much brighter classical Cepheids. If they were really brighter, then they must be farther away to appear of the same apparent magnitude. Hence the galaxy, together with all its companions, abruptly doubled in distance, and the latest estimate is about 2,200,000 light-years.

At this point we may well raise the question: how are the absolute magnitudes of the Cepheids determined? This is a question worth mentioning, because, although their remarkable association was discovered in 1912, there are still doubts over the precise value of the vital law. The trouble is that, surprising though it may seem, Cepheids are very rare stars; they are conspicuous only because they are bright, and unluckily for terrestrial astronomers there are none in the Sun's neighbourhood. Ordinary parallactic measurements are therefore out of the question, and to make matters worse their spectra are unsuitable for a straight determination of luminosity. The only means of investigation is by employing a method based on the motion of the solar system through space, which gives a steadily-increasing baseline from which to investigate parallax.[1] Unfortunately various difficulties make this, too, rather unreliable. It is for this reason that the two classes went undetected for so long, and even today there is considerable difference among astronomers as to the precise value of the luminosity. Indeed there are suggestions that the accepted law gives too high a value, and that galactic distances are correspondingly less than we suppose. But the correction can only be a minor one, and there is no prospect of any drastic reassessment of the scale of the universe.

Cepheids are not the only 'clockwork' stars; another family, closely related, are called after their prototype, RR Lyrae. RR Lyrae stars are all of about the same luminosity (85 times that of the Sun), and they are much more common than

[1] Such parallaxes are known as secular or statistical parallaxes.

Cepheids. Their constant brightness makes them very reliable distance indicators, but because they are dimmer than many Cepheids they are not detectable at such great distances. They are undoubtedly present in the Andromeda galaxy, but unfortunately we cannot make them out.

We know why Cepheids and related stars change in luminosity, though the precise mechanism is obscure. The reason is that they are literally pulsating, becoming larger and smaller in the manner of a balloon that is alternately blown up and let down. This is revealed by the Doppler shift, which indicates a surface that is regularly approaching and receding. This in turn produces a rhythmic change of surface temperature, thereby affecting the radiation and the resultant brightness.

It is now time to turn to the less predictable inhabitants of the Galaxy, the semi-regular variables. These are mostly red giants, and a fine example lies in the constellation Cetus (the Whale). Catalogued o Ceti, but more generally known as Mira ('the wonderful star'), it changes its brightness by up to a thousand times in a very rough period of 11 months. Neither its period nor its maximum and minimum magnitude are predictable, and this errancy makes it and other companions especially suitable for amateur observation.

Mira lies in a desolate part of the sky, and is therefore easy to find when near maximum; it usually stays visible to the naked eye for a few weeks at a time. Some maxima are very faint, when it rises to only the 5th magnitude; at other times, as in 1970, it has become as bright as the Pole Star (2nd magnitude). What makes it even more interesting is its tiny white dwarf companion, which because of the glare is best seen near minimum.

There are many shades of behaviour in semi-regular variables, and about the only characteristic common to all is that the rise to maximum is more abrupt than the subsequent decline. Many suffer from 'standstills' of unpredictable length, when they may remain at the same magnitude for several months or even a year at a time, and others have two distinct types of maxima, one bright and the other faint. Once again, although it is clear that the variability is due to some sort of pulsation, we have no idea just why their behaviour should be so erratic. Nature is usually highly organized, and in a way Mira-type stars are even more mysterious than the clockwork variables.

Perhaps the most famous genuine variable star in the sky is Betelgeuse, the red giant forming the left-hand shoulder of Orion (as we look at him). Ranging from the brightness of Rigel, in the Hunter's right knee, to that of Regulus (α Leonis), it bridges the gap between semi-regular and irregular variables. Betelgeuse has a very approximate period of 5 years, but it is not at all precise, and its fluctuations do not follow any standard pattern. It is remarkable for having actually been seen to vary in size. It is one of the largest known stars, and despite its distance of 240 light-years it presents the greatest angular diameter. This size, however, is fantastically minute; if we removed a new penny to a distance of 70 miles, it would appear the same diameter as Betelgeuse does from the Earth.

It is perhaps misleading to say that it has been 'seen' to pulsate. No telescope yet constructed, and none likely to be, can hope to show so tiny a disk, but it is possible to modify the instrument and produce a false disk. This device, known as an interferometer, was used on the 100-inch telescope at Mount Wilson Observatory to measure the diameters of Betelgeuse and some other giant stars, and discrepancies between the results are attributed to genuine changes of size.

Antares is another red giant to vary irregularly, though its changes are not very apparent from British latitudes. More interesting is γ Cassiopeiae, which varies erratically between magnitudes 2 and 3½ (in 1936 it suddenly outshone the Pole Star), and the presence of two neighbours of approximately its mean brightness make the fluctuations very easy to trace. This is less true in the case of Betelgeuse, for instance, and in addition its orange tint makes comparison still more difficult. Unlike semi-regular variables, irregular variables are sometimes ordinary white stars, though with peculiar spectra. The erratic performer in Cassiopeiae is white.

Some irregular variables have been really startling. The most remarkable of all, η Carinae, lies in the far southern sky and so is never visible from British latitudes. Originally of the 4th magnitude, it began to brighten in 1816, flickering irregularly until by 1840 it was second only to Sirius! It held the challenge for 9 years and then faded slowly; by 1870 it was lost from naked-eye visibility, and by 1885 it had reached its present level as a dim telescopic object. It may be significant that η Carinae is immersed in a cloud of interstellar gas, and it is possible that the star will start to flare up again at any moment.

The explosive mood of η Carinae once more bridges a gap between classes of stars, and its violence leads us to the most drastic objects in the universe: novae and supernovae.

CHAPTER 21

Exploding Stars

IN THE early hours of the morning of February 22nd, 1901, an amateur astronomer named T. D. Anderson was taking a stroll after some hours of meteor observing. Glancing up at the western sky he noticed something curious about Perseus; Algol, the Demon Star, had turned on an extra piece of black magic by conjuring up a companion. Where there had been no star the night before, nor, indeed, for centuries before, a bright newcomer was now rivalling its long-established neighbour. By the following night it was brighter than the nearby Capella, and still rising. But its reign of glory was brief. In less than a week it had started to dim; for three months it hovered on the fringe of naked-eye visibility, and in September it disappeared. With the aid of a telescope we can still see the faint ghost of a stellar catastrophe that would make the searing Sun seem like a firefly: an exploding star, or nova.

Nova means 'new', but this is not at all true; if we examine early photographs of a region where a nova has subsequently appeared, an inconspicuous and apparently innocuous star is always found occupying its precise position. Then in the space of just a few hours it suffers a titanic burst of radiation that raises its luminosity by from 50,000 to 100,000 times, and, if it is a nearby star, flashes it into brilliant visibility. There is no way of telling which stars are likely to explode, and so novae cannot be predicted; most of the bright ones have been discovered by amateurs, who have more time and inclination to look at the sky than professional astronomers.

Altogether there have been 25 naked-eye novae this century so that the number in the fainter category runs into dozens. They are detected by observatories who run 'patrol plate' programmes throughout the year, photographing the sky with special cameras to check up on novae and unexpected comets. Most novae, near or far, show the familiar characteristics of a phenomenally rapid rise to maximum followed by a slow decline that may last months or even years. This is well shown by the light curve of Nova Persei, the one discovered by Anderson.

Just why a star should explode in this unpredictable manner is a major mystery, and most of our knowledge of novae is concerned with the 'what' of the matter rather than the 'why'. The spectroscope has revealed that the star does not literally disintegrate; instead, it flings off a vast envelope of gas that may eventually become large enough to be visible

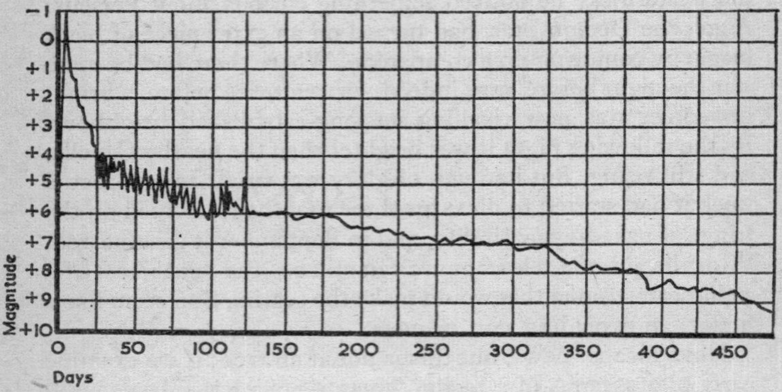

FIG. 48. *Nova Persei, 1901.* Notice the rhythmic flutterings soon after the decline. These probably represent enormous pulsations in the wrecked star.

with a telescope. Indeed Nova Persei is now surrounded by a luminous shell that is still expanding, although it takes years to detect the movement. Nevertheless this is furious activity on the astronomical time-scale, and it is interesting to compare early and recent photographs of this particular object (Plate IV).

Because novae are always so unexpected we know very little about the state of the star before its sudden surge; by the time it is detected it has already reached maximum brilliancy. We were luckier in the case of a nova which appeared in Aquila in 1918, for it was caught early in the rise and proved to be an A star, about the size and mass of the Sun but considerably hotter. What is more, early photographs, exposed for some other purpose but chancing to include the star in their field, have shown that in its pre-nova state it was slightly variable. It has now dimmed down to its original brightness, and curiously enough is still variable. It is almost as if these outbursts might be a part of its life pattern.

Evidence that this type of behaviour is not impossible is afforded by the remarkable star T Coronae Borealis, often referred to as the Blaze Star. Originally a faint object of the 9th magnitude, it shot up to 2nd magnitude on May 12th, 1866, fell away rapidly (it was visible with the naked eye for only eight days), and returned to its original brightness. Eighty years later, on February 8th, 1946, it again shone out. It is now once more a faint object, but the Blaze Star may have future surprises in store; the main fascination of the sky is that we can never trust to routine.

Some novae are more lethargic. Nova Herculis appeared in December 1934 and stayed near maximum until March of the following year, and another object, Nova Pictoris, is still dimming after its eruption in 1925. In this case we ought perhaps to include the erratic η Carinae, which some astronomers do indeed class as a pseudo-nova. Among recent naked-eye novae we may cite the case of Nova Delphini 1967, which had faded by only four magnitudes at the end of 1970.

Just occasionally, perhaps once in 300 years, we may expect a really brilliant star to blaze out; the last three occurred in 1054, 1572, and 1604, while the one which appeared in Scorpio in 134 BC is said to have inspired Hipparchus to draw up a chart of the sky. Originally these were thought to be ordinary novae, appearing bright because they were close, but we now know better because, oddly enough, of what happened in the Andromeda galaxy in 1885. In that year a star shone out among the nebulosity, and because it just reached naked-eye brilliance it was not much fainter than the combined light of the rest of the galaxy. However, it was not then realized that the Andromeda object *was* an external galaxy. It was supposed to be a nebulous cluster of stars inside our own stellar system, and as such the nova was nothing unusual.

But in 1925, when Hubble first realized its true nature, the rôle of the nova became much more remarkable. To appear so bright at such a colossal distance, the eruption must have been enormous compared with even an ordinary nova – something like 1,000 times as violent! Clearly, here was an entirely different class of exploding star, a supernova. Since that time many ordinary novae have been observed in the Andromeda galaxy, but never another supernova. They are obviously very rare, and it explains why those observed in our own galaxy have occurred at such extended intervals.

The supernova observed by Chinese astronomers in 1054 leaves a legacy in the form of the Crab Nebula, a faint smudge of light in Taurus. It can be picked up with a small telescope, but it takes the camera to reveal its true form: a grotesque mass of gas whose contours bear witness to the unimaginable fury of its birth; Hoyle suggests it was equivalent to the simultaneous explosion of 1,000,000,000,000,000,000,000,000 hydrogen bombs. Its present width is about 500 times the diameter of Pluto's orbit, and it is still expanding at the unimaginable rate of 3,000,000 mph.

It was not possible directly to relate the Crab Nebula with the Chinese observation, for the early astronomers were necessarily imprecise in locating the position of a star – although, considering their difficulties, their observations were amazingly accurate. So astronomers had to do some detective work. By taking photographs as far apart in time as possible and measuring the increase in the cloud's extent, it was possible to backtrack and decide when all the material had been concentrated in one spot, at the beginning of the explosion. The results agreed excellently, and there can be no doubt that the two objects are one and the same.

The Crab Nebula is about 5,000 light-years away. This means that the explosion did not occur in 1054 at all; it took place long before the Chaldean shepherd-astronomers began to watch the sky, and all that time the first rays were flashing their message through space. In short, our knowledge of the Crab Nebula is, and always will be, 5,000 years out of date; and the further we look into space, the more antique our knowledge becomes. The supernova observed in the Andromeda galaxy occurred long before recognizable men had appeared on the Earth at all! In some ways this is aggravating, but in many fields it becomes a positive advantage to be able to look thus into history. We can at least be sure that many supernovae have flared and died in the Galaxy, but there is no way of telling when we shall receive their news.

Tycho's Star, as the supernova of 1572 is called, was not actually discovered by Tycho Brahe, but his observations of it are certainly the most important that were made. He first saw it on November 11th, when it shone out in Cassiopeia as brightly as Venus, and was actually visible during the day; it maintained this brightness for three weeks and took altogether well over a year to fade from view. Tycho's magnificent instruments enabled him, with the naked eye, to assign it a position very close to where we now see a faint variable star, and indications are that this is possibly the corpse of the supernova.

In the case of Kepler's Star of 1604, which appeared in Ophiuchus, no accurate positions are available (Kepler, for all his mathematical gifts, was too much of a mystic to be a good observer; his main concern with the new star was in watching it 'sparkling like a diamond with prismatic tints', which is not exactly helpful). However, astronomers have discovered a strong source of radio waves in the region in which it appeared, and this has enabled the dim wreck of the supernova to be identified. It is no coincidence that the Crab Nebula is one of the most powerful known radio objects in the entire sky, and one of the first to be detected by the new science of radio astronomy which developed so suddenly after the last war.[1]

We might compare a star and a supernova with the Calder Hall reactor and a hydrogen bomb. Both have roughly the same energy potential, but whereas the bomb releases it all in a fraction of a second, the slow reactor maintains a much lower output almost indefinitely. Why a star, which is essentially a slow nuclear furnace, should suddenly commit suicide in a gigantic outburst of energy, is something that we do not yet understand; but some astronomers believe it may be the normal climax, and a fitting one, to its life.

[1] Tycho's Star has also proved to be a radio source.

CHAPTER 22

Star Clusters and Nebulae

THE GALAXY contains 100,000,000,000 brothers of the Sun, and most, but not all, are scattered at random throughout its catherine-wheel structure. Here and there, however, we find a group of stars relatively close to each other, numbering anything up to several hundred. About 300 such 'open' clusters are known, and the Galaxy probably contains several thousand altogether.

There is also a completely different type of stellar aggregation known as a 'globular' cluster. These are comparatively rare; only 100 are known, and there is good reason to suppose that few remain to be found. But what they lack in numbers they more than make up for in content, for they each contain literally hundreds of thousands of stars packed together (relatively speaking) like a swarm of bees. The inhabitant of a planet revolving around one of these suns would enjoy a night sky of unimaginable brilliance.

The most obvious of the open clusters is the group known as the Pleiades, or Seven Sisters. Seven stars are visible with the unaided eye on a clear night, but a small telescope will reveal dozens; the Sisters have many relatives! Altogether there are about 250 Pleiads, but the cluster is not a particularly important one; it appears bright and large only because it is a mere 400 light-years away, and because its leading stars are very luminous. Other unrelated stars naturally appear in the same line of sight, either nearer or more distant, but it is easy to weed out these intruders. For the group as a whole moves through space as one unit, and the proper motions of the individual stars must therefore correspond with each other.

A chart of the brighter Pleiads is shown in Fig. 49. The seven sisters themselves have all been named, together with their parents, Atlas and Pleione, and it was the star Merope which began a tremendous conflict in the astronomical world in 1859. In October of that year the French astronomer Tempel, better known for his comet discoveries, announced that the star was the centre of a faint nebula (a cloud of glowing gas), whose intensity he compared to a stain of breath upon a mirror. Other observers were unable to see the Merope nebula at all, while a few recorded it as being much more extensive, reaching the nearby stars Electra and Celaeno. Argument

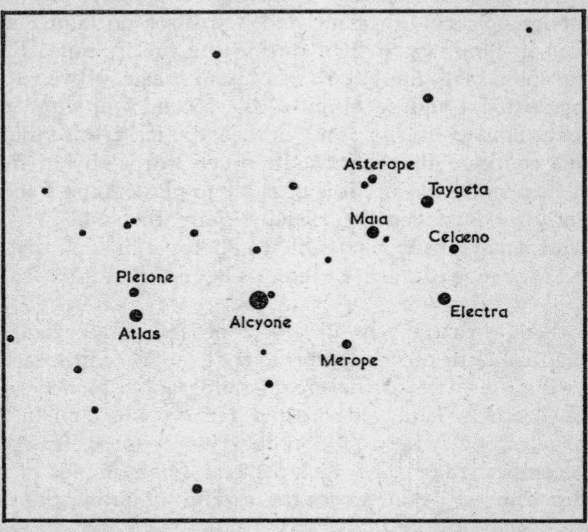

FIG. 49. *The Pleiades.* It is interesting to compare this binocular view with the photograph in Plate VI.

raged intermittently until 1885, the year in which photographic pressure was brought to bear.

It is in stellar astronomy in particular that the introduction of photography, allied to spectroscopy, has produced such shattering advances in our knowledge; it is safe to say that had the sensitive plate remained undiscovered we would now know little more about the stars than we did in 1880. Photography has extended our horizon a thousandfold; it has made it possible to analyse the composition of a star invisible to the eye in the telescope used to take the photograph; it has added millions of stars to the Galaxy and millions of galaxies to the universe. It is worth using the Pleiades as an example of how powerful a tool the photographic medium has proved itself.

Photography as a science was instituted in 1839, and even during its primitive 'wet-plate' stage it was realized that the photographic emulsion contains an enormous advantage over the human eye: it does not work instantaneously. Show the eye a telescopic field of stars, and it will see no fainter stars after a half-hour watch than during the first minute. But a photographic plate literally builds up an image. If we expose two plates for 5 and 30 minutes, the second will show stars 6 times as faint as the first; and, in theory at least, this building up will continue indefinitely. In other words, by giving a sufficiently long exposure it is possible to photograph a star or nebula that must remain visually quite invisible. With a relatively small lens it is possible to photograph stars that are beyond the range of the eye even in a large observatory instrument.

By 1880, with the new dry-plate process established, the time had come to put this enormous advantage into practice. It was done by two of the great photographic pioneers, the French brothers Paul and Prosper Henry, who had built a special telescope at the Meudon observatory for the purpose. In December 1885 they took several photographs of the Pleiades using a 3-hour exposure, carefully guiding the telescope to follow the stars, and the results were extraordinary. Not only was the Merope nebula brought out in brilliant detail and extent, but the bright Pleiad Maia was shown to also have its own private veil, and further efforts showed the entire cluster to be wreathed in nebulosity that was completely invisible in the largest telescopes (Plate VI). It was photography's first and perhaps most spectacular triumph.

This nebula is of course genuinely associated with the cluster, but it might perhaps be truer to say that the cluster is associated with the nebula. Most astronomers believe that stars are always formed in clusters like the Pleiades, and that these clusters themselves condense from nebulae. However, they are not stable over long periods of time; due to various external influences the stars gradually drift apart, and only in the case of recent births do we see star clusters in their original state. The Pleiades, which were formed a mere 6,000,000 years ago (an incredibly short time), have not yet gained their independence, and the faint nebula is the remains of the primordial cloud that gave them birth.

Such is the makeup of the Galaxy that we can see all this history being enacted before our eyes. Stars were not all formed at the same epoch. We can still see nebulae that are in the process of giving birth to stars; we can see young clusters; finally we end up with stars like the Sun that have lost their affiliation and are living the rest of their lives on their own. Astronomy is the one science which is so open about its history. And when we look beyond the confines of our own Galaxy, the distances are so enormous that the finite velocity of light allows us to look back in time for millions of years. The exploitation of this gift is a matter for Chapter 25.

There are plenty of spectacular open clusters in the sky. The Pleiades are in Taurus, which also contains a much looser association, the Hyades, at a distance of only 130 light-years. These apparently surround α Tauri (Aldebaran), but Aldebaran is really much closer than the cluster and simply appears in the same line of sight. Another loose cluster, just visible with the naked eye but best seen with binoculars, is Praesepe (the Beehive), in Cancer.

Since open clusters are scattered throughout the Galaxy, it is natural to expect them to occur most frequently in the Milky Way, which is simply our view through the greatest depth of stars. This is indeed so, and we find the most notable clusters in the grand constellation Perseus. There are two of them, side by side, and they are physically connected; with a small telescope they are a most magnificent sight, the very background of the night sky being strewn with faint stars, while the brighter suns are concentrated around two nuclei. This, the famous Double Cluster of Perseus, is visible with the naked eye as a strong concentration in the Milky Way. Telescopic clusters abound in this part of the sky.

It is obvious that if we are very close to a star cluster the stars composing it will appear more spread out. Thus, although it is not immediately apparent, many stars in the constellation Orion and five of the seven bright stars in Ursa Major are genuine clusters. It is of course pure chance that the ancients so conveniently included them together in constellations; and in any case Orion has many stragglers in outside regions, while the brilliant star Sirius actually belongs to the Ursa Major group. This is made clear by studying the proper motions of the various stars. In some cases, these motions show that the stars are radiating from a common point, as though from an explosion. For instance, a study of 17 stars near ζ Persei suggests that they all formed a compact group a mere 1,300,000 years ago – an astonishingly short time on the cosmic scale. Most of the time, however, it is very unsafe to suppose that stars apparently near each other are really associated; distances and absolute magnitudes vary so considerably that the only way to prove an association is by their motions. For instance, the bright stars Castor and Pollux in Gemini have nothing to do with each other.

Mention of Orion leads us to the true nebular department. Just below the Hunter's belt the unaided eye perceives a glow of light, the famous Great Nebula, which in fact is only the nucleus of a colossal cloud of interstellar gas that extends over the entire constellation. The eye alone can follow it for only a few degrees, but the camera once more traces its true extent. This is the primordial cloud responsible for Orion's stars, and in the bright nucleus they are still being formed: an unimaginably slow process that we see as though a moving film were suddenly 'frozen' to a single frame. The whole universe, in fact, is a turbulent mass of activity which we, because of our microscopic life span, have too fleeting an existence to appreciate.

There are many nebulae that we can catch in the act of giving birth to stars, though most of them are poor telescopic objects; they are sights for the camera only. They appear to be glowing, but in general this does not mean that they are exceptionally hot. In the Orion nebula, for instance, a small telescope wlll show four stars forming a rough square in the central nucleus. The group is known as the Trapezium, and the surrounding gas is made to shine by reaction with ultra-violet radiation from these stars. The effect is very like a terrestrial aurora, though on a colossal scale, for the gas, apparently dense, is really far thinner than the best vacuum we have been able to produce artificially.

Elsewhere in the Milky Way we find nebulae on a much smaller scale. There is a good example in Lyra, near the bright star Vega; it is known as the Ring Nebula because it looks exactly like a celestial smoke-ring. Large telescopes reveal a faint star at its centre, and it seems very probable that in ages past the gas was expelled from the star, possibly after a nova-type explosion.

The fact that most of these gas-clouds are essentially non-luminous is brought home to us by the many dark nebulae that are distributed throughout the Galaxy. In this case we can see them only in outline, by the absence of the stars they obscure from view; Herschel was the first to notice what were apparently holes through the star layers, but it needed photography to explain their true nature. By taking small-scale photographs of different regions of the Milky Way, thereby showing a large area at a time, the famous astronomer Barnard was able to show their curious configurations. Basically they do not differ from the bright nebulae, but it just happens that there are no nearby stars to excite them to luminosity.

Some of these dark nebulae are in fact near enough to be visible without a telescope; the most famous is where the Milky Way passes through Crux, the Southern Cross. Known appropriately as the Coalsack, it appears as a great black patch in the shining river of stars. The Coalsack is only 400 light-years from the Sun, and is one of the closest of the dark nebulae, but they exist all along the course of the Milky Way. But for their obscuring effect, indeed, our night sky would probably be brighter than when lit up by a Full Moon! The photographs by Barnard, taken during the first years of the present century, are among the best ever secured of the hoards of bright and dark nebulae that cover the sky.

In a way it is misleading to speak of definite 'clouds', since space is filled with an extraordinarily dilute fog of hydrogen gas and dust particles, and the nebulae are little more than huge local concentrations in the general murk. Their intensity is deceptive, for they are incredibly tenuous and it takes several light-years of thickness to dim the starlight even appreciably. By comparison even a comet's tail is a rugged object.

At this juncture it is worth mentioning the systems by which clusters and nebulae are classified. The classical list, containing 105 objects, was compiled by Messier, an eighteenth-century comet hunter. He, together with some other fellow-workers, was constantly embarrassed by sighting a cometary object that on closer inspection proved to be a faint nebula or condensed star cluster; consequently he drew up his list of 'false' comets. The curious sequel is that despite his brilliant success (Messier discovered altogether 13 comets) he is now best remembered for his catalogue! The objects listed are designated by the letter M, and conveniently refer to most of the brighter northern features, so that they form a useful guide for an observer with a small telescope.

The next and most important list was drawn up by Herschel himself. It was naturally far more extensive than Messier's; altogether he listed 2,500 clusters and nebulae, and he subdivided them into 8 distinct classes, according to their nature, indicated by a Roman number after the letter H. Thus we have H.VI.33 and H.VI.34 as the two clusters forming the Perseus Sword Handle. There are also many more modern catalogues, such as the New General Catalogue of 1888 (NGC), but Messier and Herschel between them covered all the brighter northern objects, and in any case the NGC is only an extension of the Hanoverian astronomer's work.

Most nebulae, light or dark, are confined to the main plane of the Galaxy as though embedded in a gramophone record, but the globular clusters are an entirely different matter. They are concentrated chiefly at the galactic centre, which from our point of view lies in the direction of the constellation Sagittarius, the Archer, and this determined independence raises questions of its own.

Three only are visible with the unaided eye, and of these two, ω Centauri and 47 Tucanae, are southern objects. They are both magnificent sights, but owe their brilliance to their nearness. By contrast the brightest northern globular, M.13 in Hercules, is little more than a dim patch of light, despite the fact that it probably contains 100,000 stars as bright as the Sun.

M.13 has a diameter of about 300 light-years, and were it as close as the Pleiades it would be a magnificent sight indeed: a blazing ball of stars 30° across! But in fact it is very remote, and to gauge its distance of 36,000 light-years we have to call on the faithful RR Lyrae variables. Globular clusters almost always contain several of these stars, but never any Cepheids. This is the preferable alternative, for RR Lyrae stars are rather more reliable distance indicators.

Globular clusters are permanent star-cities, unlike the open clusters which are diffusing their members into space. The stars, in fact, are revolving around the cluster's centre, and have been doing so since their birth. Unfortunately we have no clues as to why these gigantic swarms of stars should have been formed. The Galaxy, for once, lets us down, and nowhere does there seem to be a globular cluster actually in the process of formation. Strangest of all, there are no nebulae associated with them. The spaces between the myriad stars have somehow been swept free of all primordial material.

Some astronomers have gone so far as to suggest that the globular clusters are adopted infants, and were formed independently of the rest of the stars in the Galaxy. This claim is rather hard to substantiate, but they are certainly the most inexplicable members of our star system. Let us now stand back and try to assemble stars, clusters, and nebulae into their proper order, and see what the Galaxy as a whole looks like.

CHAPTER 23

The Milky Way

SUPPOSE WE lived in a world where there were neither mirrors nor cameras. How could anyone find out what he looked like? The answer would be to feel the contours of his face and to associate what he felt with what he saw on other people's faces. In this way he could at least establish that he had the full complement of eyes, ears, etc., and by some intelligent guessing he might be able to build up an even more comprehensive picture. If this takes us back to the Greek philosophers, it is perhaps surprising to find it a fairly accurate analogy of the problem of probing the form of the Galaxy.

In fact the trouble is precisely that we cannot, as the previous page hopefully suggested, 'stand back' and see what the Galaxy looks like. We are inside it, firmly embedded in it. We are in the position of someone standing in one spot in Hyde Park and trying to plot its perimeter. Here and there he can perhaps glimpse the edge, but in most directions trees obscure the view after a few hundred yards. What lies beyond is mainly guesswork. In the Galaxy the 'trees' are the nebulae, mostly dark, which limit our view to a few hundred or thousand light-years. Without these obstructions, and assuming that space were perfectly transparent, the matter would be a straightforward one.

However we must make the best of it, and the first person to put forward a constructive theory based on observation was, of course, Herschel. What he did was to count the number of stars of different apparent magnitude in selected, scattered areas of the sky. Assuming that they were all of roughly the same absolute magnitude, their brightness was a factor of distance; it was accordingly possible to decide how far he was seeing into space in different directions. For example, when he pointed his telescope at the band of the Milky Way he was plumbing far greater depths than when he looked at right angles to its plane.

Herschel's basic assumption was of course wildly wrong, for stellar luminosities differ enormously, but he did at least prove to his own satisfaction that the Galaxy was in the form of a thin disk with an irregular, roughly elliptical outline (Fig. 50). The Sun was buried almost half-way in its thickness, which explained the Milky Way effect when we looked through the full stratum of stars. The curious rifts are due to dark nebulae, which Herschel and his contemporaries imagined to be true vacuities.

Herschel's Galaxy was a static stellar system, without rota-

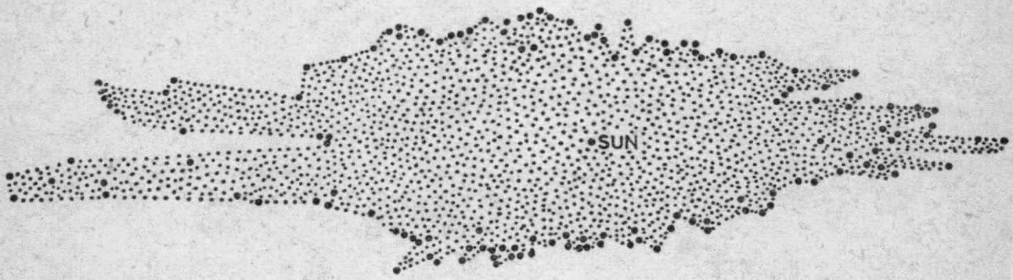

FIG. 50. *Herschel's model of the Galaxy.* This is his cross-section through the thin 'slab' of stars. He assumed an even distribution of stars within its boundaries, and was forced to introduce considerable indentations and projections in its outline to account for regional variations in the night-sky density.

tion, and to twentieth-century eyes it has an awkwardly artificial aspect; yet, in many respects, it was much closer to reality than anything proposed during the following century. He was, for instance, prepared to believe that the Galaxy was but one of many that made up the universe, whereas as recently as 1920 it was being maintained that if any other star cities existed they were too distant and faint to be seen with any telescope; and in 1900 it was claimed that the diameter of the universe itself was a mere 20,000 light-years (compared with the present estimate of the width of the Galaxy at 80,000 light-years!). The great jump clearly came in very recent times. In fact it followed immediately from Hubble's discovery, in 1925, that the 'nebula' in Andromeda was not a cloud of gas at all, but an immensely distant external galaxy. Here was our mirror-image; here, presumably, was how our own system ought to look. And armed with this helpful knowledge, observations began to slip into place, just as a half-finished jigsaw makes sense when one can see the finished picture.

The present concept of the Galaxy is of a slowly-rotating spiral, whose elevation and plan is shown in Fig. 51. At the

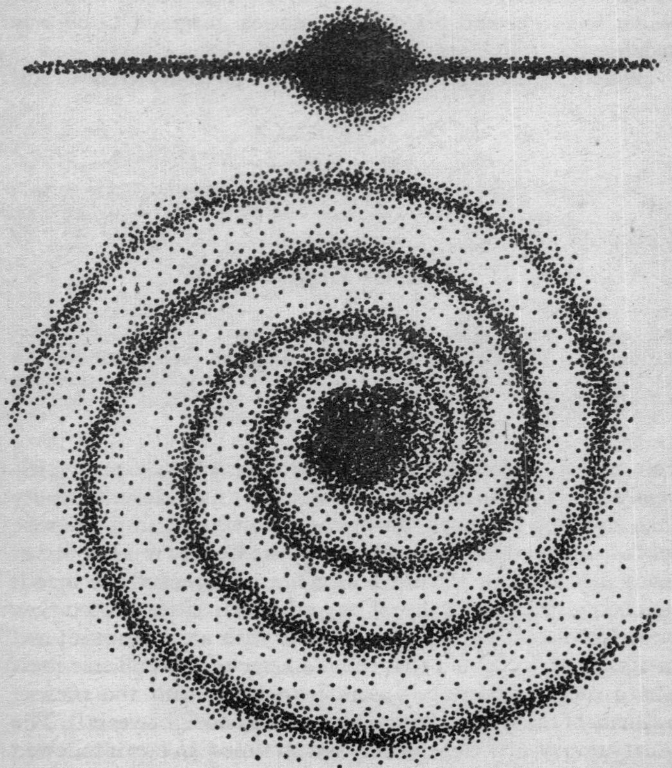

FIG. 51. *Two views of the Galaxy.* The spiral structure is nothing like so neat as the one drawn here, but its overall nature is clear enough.

centre of the system the stars are grouped into a roughly spherical core about 20,000 light-years across, while from this, in what we might call the equatorial plane, trail two huge arms of stars laced with gas, extending some 30,000 light-years away from the nucleus. Because of the galactic rotation these arms are curved in a spiral fashion, dragging so far behind that they are partly intertwined. The Sun, so important to us and so insignificant in this immense population, is situated in one of the arms, rather more than half-way from the centre to the edge; once again a comedown from 40 years ago, when we believed ourselves to be at the very centre of the system. To be precise, we are 27,000 light-years from the nucleus but within 100 light-years of the Galaxy's central plane.

The Galaxy as a whole is rotating – but it is not rotating as a whole! If it were a rigid structure the 'cosmic year', or period of revolution of a star around the nucleus, would be the same at all distances, just as every point on a wheel completes one revolution in the same time. But as things turn out the 'year' becomes longer with increasing distance. It is about 100,000,000 terrestrial years at a distance of 10,000 light-years from the centre, while the outermost reaches of the arms take at least four times as long. The Sun's cosmic year is of the order of 220,000,000 years.

We might at first be tempted to compare these figures with the planets in the solar system; the farther from the Sun, the longer the period, in obedience to Kepler's third law. But the Galactic set-up is more complicated. In the case of the solar system the Sun contains the vast preponderance of the total mass, but in the Galaxy the distribution of stars is much less partisan, and we cannot apply mathematical rules until we know far more about the make-up of the nucleus and the arms. All we can do is observe stellar movements and decide from them how the rotation varies from zone to zone; and in most cases we are baulked by the astronomer's bane, interstellar dust and dark nebulae. From our position, at the back of the stalls, we cannot see farther than about 7,000 light-years towards the nucleus. This leaves a colossal gap of 20,000 light-years virtually unknown, and we are reduced more or less to guesswork.

If we could only tear free from the thin plane of the Galaxy we could see the whole structure laid out before us.

The arms are thin and extended for rather the same reason that Saturn's ring is so attenuated. Of course the Galactic case is far less extreme; the stars have not yet been organized into well-behaved groups, and they have several light-years of freeway on either side of the plane. The interesting thing is that the obscuring gas and dust is more strictly regimented into a disk-like structure, and this has significant personal consequences. The Sun itself is moving on its colossal journey in an orbit that is not precisely parallel with the Galactic plane. At the moment we are almost exactly in it, but there is every reason to believe that a million years ago we were a few hundred light-years to one side. At that epoch we would therefore have been clear of the dark nebulae, or at least the majority of them, and could probably have seen the true nucleus of the Galaxy; this awesome sight was removed as soon as man's first ancestors began to stir on the Earth's surface.[1] One day in the far future we shall move out of the cosmic shadow, but it is doubtful whether human eyes will be here to watch the slow emersion of the brilliant nuclear star clouds.

One of the most important features of the Galaxy is the great difference between the populations of the nucleus and the spiral arms; the stars to be found in these regions can be divided into two distinct classes, Population I (arms) and Population II (nucleus). Population II stars are mostly red giants, with cool surfaces and huge volumes, while the Sun, together with its main-sequence companions, belongs to Population I.

It has been known for a long time that the Galaxy consists of these two major groups of stars, but at first they were distinguished not by their physical nature but by their movements, which were either fast or slow. Unlike the low-velocity stars, which belonged to the spiral arms, the high-velocity objects all moved in small orbits around the nucleus that were inclined at all possible angles, so that they effectively built up the central sphere of stars. We could not directly observe the nucleus because of obscuring matter, but there were some high-velocity

[1] Coincidence?

stars whose eccentric orbits had carried them within view, and these traitors gave valuable information about their hidden companions. From their number it was possible to calculate that these stars represented the bulk of the Galaxy's population.

Not until 1943 was the nature of these high-velocity objects settled, and it was done not by peering through our murky dust clouds, but by gazing through the transparency of intergalactic space to the Andromeda galaxy. A study of this noble object had proved without doubt its general similarity to our own system, with its spiral arms and blazing nucleus, and careful photography had actually revealed the myriad stars that thronged the arms themselves. But the nucleus remained inscrutable; the stars were so densely packed that they simply registered as a mist of light. What was needed was a more sensitive photographic plate to pick out the most luminous stars in this throng.

Walter Baade, the great German astronomer whose recent death was a tragic loss to galactic astronomy, realized this need. Photographic plates are normally sensitive to a wide range of colour; usually, in fact, it is wider than that visible to the human eye. But it is possible to narrow their sensitivity into a much more restricted range of wavelengths. For example, blue-sensitive and red-sensitive plates are commonly used in astronomical research.

Baade, in his efforts to resolve the Andromeda nucleus into stars, had used blue-sensitive plates; but he realized that if the stars were red, a red-sensitive plate would naturally give better results. So he used the 100-inch telescope at Mount Wilson Observatory to take photographs of the nucleus in red light (making full use of the war-time blackout!), and succeeded in resolving it into stars. This spectacular result meant that the stars must be red giants, and he called them Population II stars, to distinguish them from the much hotter and whiter stars in the spiral arms. Later work has shown that similar galaxies also have these two distinct populations. Population II stars are the high-velocity objects that we can only glimpse in our own galaxy.

Of course, red giants are not confined to the nucleus; Betelgeuse and Antares are both relatively near, and there are many more besides. This is because they are also scattered in the empty lanes between the great lassoes of Population I stars, in the dust-free regions of the Galactic wastes. This is one very suggestive feature. There is no dust in the nucleus, despite the vast tracts farther out, and if we are to believe that the solar system condensed out of material dragged into orbit by the Sun's attraction, it follows that there can be no planetary systems near the centre of the Galaxy. In any case, a red giant would hardly make the most hospitable parent.

These stellar controlled-zones take us back to the globular clusters, which consist very largely of red giants, and we see now why they are concentrated near the nucleus. In fact they form their own private halo around the nucleus. They travel in fast, eccentric orbits that are just like those of ordinary Population II stars, only considerably larger. This means that by the laws of chance a good proportion are to either side of the obscuration belt, so that we can see fully a half of all that are likely to exist (Fig. 52). The open clusters are a very different proposition, for they are formed from Population I stars and so are confined to the plane of the spiral arms; in their case a great many must be hidden.

Much of our recent knowledge of the Galaxy has accrued from radio investigations, and the sensational success of the radio telescope in probing for data comes about through two reasons. First, radio waves are hardly affected at all by interstellar matter, so that they can 'see' through the infuriating nebulae as though they were not there at all. Another and more significant fact is that objects which are intense radio transmitters are often obscure visually. The outer reaches of the solar corona are a case in point, but all over the Galaxy we find what are known as 'discrete sources': centres of emission which usually turn out to be relatively small nebulae. For example, the most powerful source in the sky, labelled Cassiopeia A, was discovered in 1948, but not until 1951 could a visual object be photographed in its position – and it was faint even in the 200-inch telescope, the largest in the world! The

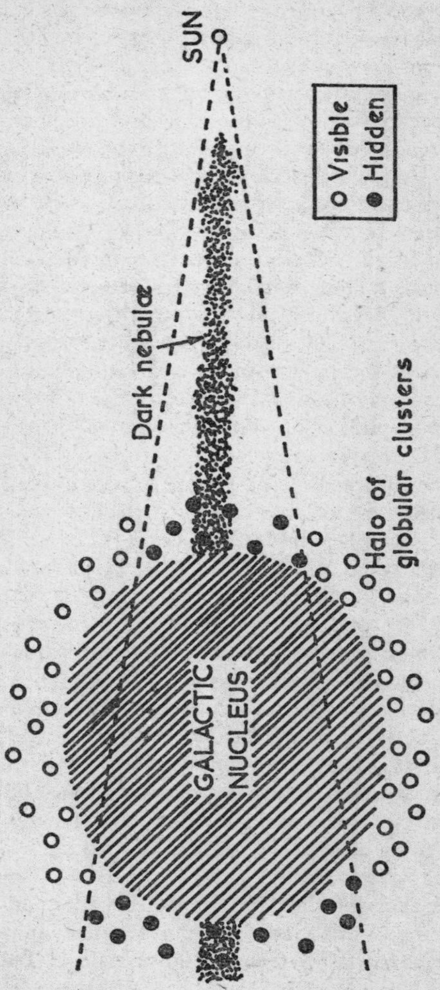

FIG. 52. *The halo of globular clusters.* Plate V (*b*) shows a similar halo of globular clusters surrounding the nucleus of the galaxy NGC 4594.

much brighter Crab Nebula is another example, and many astronomers believe Cassiopeia A to be the remains of another supernova; possibly the one observed by the diligent Chinese in AD 369.

Other investigations have shown the nucleus of the Galaxy to be very lively on radio wavelengths, and the entire star system is embedded in a colossal halo of active particles known as the Galactic corona. These are regions of which optical astronomers must always remain ignorant, just as radio workers, were they blindfolded, could know nothing of any of the individual night-sky stars, for despite their visual brightness they emit radio waves too feeble to be detectable. The two departments must always work in sympathy, and there is no question of one trying to outdistance the other.

Perhaps the most famous and far-reaching achievement of radio astronomers is the detection of hydrogen gas in the Galaxy's spiral arms. The emission we have so far mentioned comes from atoms in a highly disturbed state, while the corona itself consists mainly of electrons, which make up only half of the basic atomic structure.[1] In its normal state an atom is almost perfectly stable and does not emit any energy. But in 1945 the Dutch physicist van de Hulst pointed out that a hydrogen atom, if left long enough, would periodically undergo a slight transformation and emit a minute amount of energy that should be detectable in radio telescopes at a wavelength of 21 centimetres. The condition 'if left long enough' is perhaps an understatement; the emission from a single atom will occur for a tiny fraction of a second at intervals of 11,000,000 years! But van de Hulst believed that there was so much hydrogen in the Galaxy that the total radiation should be detectable, and the fact that this was achieved in 1951 is as good an illustration as anything of the truly vast scale of things with which astronomers have to deal.

The arms of the Galaxy are soaked in hydrogen, and by observing at this vital wavelength astronomers have been able to fully confirm the Galaxy's spiral nature. There are evidently

[1] This is really a gross over-simplification, since physicists now recognize at least 32 basic particles!

two arms, which intertwine and rift in a bewildering fashion; the problem of analysing the observations is still so complex that the picture is not yet entirely satisfactory. But without this neutral hydrogen emission the investigation would be far more complicated, and probably insoluble, since the arms themselves are mostly hidden from view. Radio astronomy has made truly sensational inroads into the Galaxy's mysteries.

It may appear at first sight that the stars play out the rôle of the Galaxy. From a long-term view this is certainly true – what is a galaxy but the stars it contains? – yet its lifetime is to be measured by its nebular rather than its stellar content. It is from the nebulae that its stars are born, and in the beginning the Galaxy was probably a spinning mass of gas: a hugely larger version of an infant star. But at the present moment its total mass is predominantly that of its stars; its interstellar resources are dwindling, and since the nuclear region is now entirely devoid of gas it is only among the Population I stars in the arms that new stars can yet be born. In this sense the Galaxy has lost its youth, for it has produced more stars than it can produce in the future; it is over half-way to death, or, if we prefer the term, bankruptcy.

There are two ways in which we can try to probe the age of the Galaxy. One is to observe its general features and decide how permanent they are; the other is to calculate the lifetimes of its stars.

Stars are being formed all the time. A Wolf–Rayet type, for instance, can exist in its blinding brilliance for only a few million years, a period over which the sedate Sun, which uses its nuclear fuel more moderately, turns out far less radiation with rock steadiness. Its age, which is of the order of 5,000,000,000 years, is evidently much greater than that of a Wolf–Rayet star. The Galaxy must therefore be at least 5,000,000,000 years old. Very little is known about stellar ages in general, so we must attack the problem from another angle to find an upper age limit.[1]

Let us return to the question of the flattening of the spiral

[1] It does, however, appear that red giants are rather older than main-sequence stars such as the Sun.

arms. In Saturn's ring we see the end product, or virtually so, of the process where by mutual collisions the particles forming the rings have regimented themselves into circular orbits almost precisely in a common plane. The same sort of process, although the agency here is attraction, not collision, is taking place in the Galaxy's arms. Once in a few thousand million years two stars, moving at random, will pass sufficiently close to perturb their orbits, and the process, if continued long enough, would iron out any eccentricities and make the stars behave like well-ordered commuters. Nothing like this has happened. The stars, among themselves, are moving in different directions – although the general Galactic rotation carries them round as a whole in the same way that a crowd of people standing on a large turntable can walk among each other but still rotate. There is also the allied fact that despite the greater rotational speed near the nucleus, the arms have not yet become wound up out of all recognition. On these grounds it is unlikely that the Galaxy has been existing, in its present form, for more than 2,000,000,000,000 years.

This gives an uncomfortable range of age, though it is obvious that the Galaxy could never have possessed sufficient material to nourish its present numbers for so long a period. What other slow processes can we detect? The most spectacular is the slow disruption of the open clusters. If we are to believe that a great many, perhaps even the majority, of stars were originally formed in clusters that subsequently broke up, we can calculate how many clusters must altogether have existed. Knowing their average lifetime, this in turn leads to a value for the age of the Galaxy of 10,000,000,000 years. In general, astronomers prefer to reduce this to about 8,000,000,000 years; and when we remember that the age of the Earth is probably not less than 4,500,000,000 years it seems possible that the planet we live on is truly 'half as old as time'.

CHAPTER 24

Galaxies and Galaxies

IN 1810, in a rare moment of self-revelation, Herschel said: 'I have looked farther into space than ever human being did before me; I have observed stars, of which the light, it can be proved, must take two millions of years to reach this earth.' He was speaking of the 1,500 'island universes' he had discovered, external galaxies that a century later astronomers were busy disproving, and 50 years later still were speaking of in terms of millions of millions. Herschel, the humble observer, was a remarkable prophet.

It is only fair to point out that before his death in 1822 he himself had ceased to believe in his island universes. The reason was this. When he pointed his telescope clear of the Milky Way, faint galaxies crowded into view. But when he looked near and through the celestial river, they faded out. Accordingly, he ultimately decided that stars and galaxies were somehow related. A relation of incompatibility, in fact; there were stars and there were galaxies, but when stars became too numerous galaxies disappeared almost entirely.

We now know the true reason for this apparent connexion. The Galaxy, in addition to its dark nebulae, is shrouded in celestial fog in the same way that a city creates an aura of murk. Once again this haze, consisting of hydrogen gas and small solid particles, is inconceivably thin, with a density much less than a millionth that of our atmosphere. But its enormous depth makes it an efficient light absorber. Therefore, although galaxies are distributed in all directions, those that we see near the plane of the Galaxy are effectively dimmed out of existence. Herschel's proof of their connexion with our system is turned, in the light of present-day knowledge, to proof of their utter remoteness. Incidentally, this attenuation is one way of measuring the light absorption by interstellar particles.

Even ignoring this superficial rifting into two hemispheres, we still find the massing of galaxies far from uniform. In some regions of the sky they crowd even closer than in other parts; there is a spot in the constellation Coma Berenices where the 200-inch telescope has photographed 500 galaxies within an area of the sky equivalent to that of the Moon! In fact from our point of view galaxies are more numerous than stars; only a few hundred million stars can be seen, whereas over the whole sky the 200-inch could register thousands of millions of galaxies. The enormous majority of these star systems, all comparable with or even larger than our own galaxy, are so far away that they are reduced to faint spots of light, much dimmer than many of the Galactic stars through which we must photograph them.

This tendency for galaxies to collect in clusters is an important one, because it turns out that the Galaxy belongs to a group, known as the Local System, consisting of about 18 star systems, collected within a volume of space some 2,200,000 light-years across. This aggregating tendency on Nature's part is significant; does it mean that galaxies, like many of the stars within them, were formed in clusters? It certainly looks like it, and to add strength to the argument we find that the Galaxy has two smaller systems revolving round it in the manner of satellites. They are visible with the naked eye, but are stationed in the southern hemisphere; the explorer Magellan was the first to notice them, in 1519, and they are known as the Magellanic Clouds. To the eye they look just like detached portions of the Milky Way.

We are not unique in this respect. The galaxy in Andromeda (which is the most important member of the Local System) also has two attendants, and undoubtedly many others are equally privileged. The Clouds, nevertheless, remain the nearest extragalactic objects, and they are extremely important because we can examine them in such great detail. Indeed, because of this bird's-eye view we know them in some respects better than we do our own galaxy. The Large Cloud is about 35,000 light-years across and 200,000 light-years away, while the Small Cloud is rather smaller and slightly more distant. We say that they are revolving around us, but the motion is so hopelessly slow that we cannot possibly detect any movement.

At first sight they look not so much like galaxies as rather irregular star clusters; they are our first introduction to that branch of sidereal systems known as 'irregular' galaxies. If they are rotating, they are doing so very slowly; they have not yet worked up sufficient speed to form a flattened spiral system. We might consider them as representing an earlier stage of development. There are also 'elliptical' galaxies, which resemble an ordinary nucleus like our own stripped of its trailing arms, and this is apparently the third and final stage of a galaxy's career.

Both Clouds are rich in Cepheid variables; in fact they collectively contain more than we have counted in our own galaxy. This is because the Clouds contain a great many hot B stars, and very few red giants, and Cepheids are basically giant B stars. In fact it was the Cloud Cepheids which gave astronomers the first insight into the Period–Luminosity Law. It is a simple reason, but none the less worth describing.

If we wish to determine the absolute magnitude of a star we must know two things: its apparent magnitude and its distance (assuming other methods are inapplicable). Now Cepheids are awkward on two counts: they are mostly too distant for parallactic determinations, and their spectra are not like those of main-sequence stars – they actually change with the star's fluctuations – so that they cannot be used to find the absolute magnitude directly. Distance estimates are therefore certain to be to some extent dubious, and the probable error would swamp the slowly-emerging but as yet unknown period–luminosity connexion.

When we turn to the Cepheids in the Magellanic Clouds things are considerably simplified, because the 'depth' of either Cloud, compared with its distance from us, is small. In other words all the Cloud stars are at roughly the same distance. If we place two lamps side by side and look at them from a distance, we know that the one that appears brighter really *is* brighter; similarly, if one Cloud star appears brighter than another, we can be sure that its absolute magnitude (whatever it may be) also differs by the same ratio. This simplifies things greatly, because in the case of Cepheids their periods are related to apparent magnitude as well as absolute magnitude. This made the relationship much more obvious. Then, having used Galactic Cepheids to discover the actual value for the relationship, astronomers harked back to the original stars to find the distance of the Clouds! It was something of a combined effort.

Extra evidence that the Clouds are very young galaxies is afforded by the fact that they both contain a tremendous amount of gas and dust – the birthplaces of future stars. Moreover red giants are lacking. It seems that these are relatively old stars, while Population I stars, like the Sun, are the youngsters, and the Cloud stars are almost entirely Population I. Things are very different in the case of the brightest of our independent neighbours: the Andromeda galaxy (M.31), and the galaxy in the constellation Triangulum (M.33). These are both spiral systems like our own, and they are actually fairly close together in space; to us M.31 is a dim naked-eye object, but to an observer on a planet belonging to a star in M.33 it would be a brilliant spectacle. M.33 appears less elliptical than M.31, because we see it almost in plan view, and it is only about 1,500,000 light-years away.

Our galaxy marks one extreme, and M.31 and M.33 mark the other extreme, of the Local System. The other members are scattered more or less in between; they are considerably fainter, and in cosmic terms they are even less important than the Magellanic Clouds. What is more, they are either irregular or elliptical, and this means that we have all three basic types represented in one group, which is extremely interesting, because they are presumably all of about the same age (it is reasonable to suppose that they were formed together). Therefore, just as a small planet such as Mars 'ages' more quickly than the Earth, in the sense that it runs through its history more quickly, so we have evidence that galaxies too have very different life spans. Irregular galaxies seem to be the adolescents; spirals are in the prime of life, while elliptical galaxies are bankrupt and have lost their fertility. If we accept that these basic types do indeed represent an evolutionary sequence, we can sketch, very roughly, a galaxy's career. Beginning as an irregular system of young B stars, representing Population I, it slowly acquires a nucleus of long-lived red giants (Population II), which effectively clear this region of gas and dust and remain more or less stable while the Population I stars are born and die in the developing arms. As the fuel is used up the arms become wizened and finally disappear, leaving the nucleus of Population II stars forming an elliptical galaxy.

Some astronomers do not accept any such evolutionary relationship between the different types of galaxy, but most do; and by looking at photographs of the brighter galaxies it is certainly possible to trace an almost continuous change of form. There are some spirals, for instance, in which the nucleus is not elliptical but is in the form almost of a rectangle, with short arms at each end. These are known as 'barred' spirals, and may well represent the transition from an irregular to a spiral galaxy. In this respect it is possible that the Large Cloud itself is beginning to turn itself into a barred spiral (Plate V (c)), while its smaller companion is still quite formless.

If we now leave the comforting intimacy of the Local System we must take another jump upwards in scale; distances are not reckoned by the million, but by the hundred or even thousand million. Perhaps it would be as well to re-cap by introducing another scale model.

A pinhead is $\frac{1}{16}$ of an inch across. Representing the Earth by such a pinhead:

The Sun is 6 inches across and 18 yards away.

The nearest star is 3,000 miles away.

The diameter of the Galaxy is 60,000,000 miles – $\frac{2}{3}$ of the real distance from the Earth to the Sun.

Already sizes have become ridiculous. Reducing the diameter of the Galaxy to 1 foot:

M.31 is 25 feet away.

The boundary of the observable universe is over 1 mile away, in all directions.

The greatest distance to which the 200-inch telescope can photograph galaxies like our own is about 3,000,000,000 light-years. Photographs of some of these immensely distant galaxies show them as appearing as nothing more than tiny blurs of light, at first sight hardly distinguishable from the foreground stars – this gives an obvious clue to their remoteness when we remember that each one contains thousands of millions of stars. But there are many nearer galaxies that we can see in much greater detail, and in long-exposure photographs some of these are superb sights. Perhaps the most beautiful of all is M.81, situated in Ursa Major and visible in a small telescope as a misty stain on the sky. It is one of the nearer galaxies outside the Local System, at a distance of about 7,000,000 light-years. The spiral galaxy M.51 we see precisely pole-on, and so have an exceptionally good view; it has a subordinate system, and the two are linked together by one of the arms. By contrast we are almost in the equatorial plane of NGC 4594, known, not inappropriately, as the Sombrero Hat Galaxy, and this gives us a splendid view of the dust clouds in its tightly-wound arms. In this particular galaxy, and in many others that are bordering on the elliptical stage (such as M.87), we can see the globular clusters which form an outlying halo around the nucleus, just as they do in our own system. Plate V shows M.81 and NGC 4594.

M.87 is just one member of a huge cluster of galaxies lying in the constellation Virgo. There are altogether at least 500 in this particular group, while the Coma Berenices cluster contains 800 galaxies at a distance of roughly 100,000,000 light-years. We could go on indefinitely. Galaxies throng the sky in every direction, and except where their light is absorbed by local matter there are dozens and sometimes hundreds in every square degree. With these very remote systems, as against the closer ones whose individual stars can be used to provide distance estimates, we have to guess their distances from their brightness, so that results are rather uncertain.

Galaxies, in relation to their size, are packed far more closely than stars – certainly those stars in the Sun's neighbourhood – and this might lead us to expect occasional near misses or even collisions. Actually this is not necessarily so, because as we shall see presently galaxies are not moving through space at random; they are all flying away from each other. But if we consider two galaxies in a fairly dense cluster, it is still on the cards that they could collide; and the scale of the disaster would release a phenomenal amount of energy. How could we detect this?

We have already mentioned how radio telescopes discovered two sources of intense radiation in the sky which are associated with very faint objects (Cassiopeia A and the Crab Nebula). Conversely, bright stars such as Sirius are undetectable on radio wavelengths. There seems to be no compatibility between visual and radio brilliance, and it was therefore no surprise to find that the second most powerful source, located in Cygnus, could not be associated with any visual object. Known as Cygnus A, it was for many years a complete mystery.

By 1951 protracted radio observation had pinpointed the position of Cygnus A to within the finest possible limits, and it was now feasible for two astronomers, Baade and Minkowski, to examine the region very thoroughly with the 200-inch telescope. The main drawback of the radio telescope is its blurred vision, or, more technically, its poor resolving power. For instance, the naked eye can just divide two stars only 3′ apart, while one of the greatest radio telescopes in the world, the 250-foot bowl at Jodrell Bank, cannot by itself divide objects spaced less than ½° apart – and smaller instruments are still less precise. It could therefore define the position of Cygnus A to within a circle ½° across, but this was not nearly good enough for so sensitive a telescope as the Palomar instrument. It would have taken weeks to search this area; what is more, the source could have been any one of dozens of faint galaxies. So more disciplined radio methods had to be used, involving two telescopes placed several miles apart to increase the resolving power.

After most careful work Baade and Minkowski found what must be the source, a fuzzy and distended object that turned out to be nothing less than two galaxies in contact, at a distance of 700,000,000 light-years! If the phenomenon is awesome, the distance is incredible; an object so remote as to be visible only with the largest telescopes is the second most powerful source in the sky. This was clearly a most gratifying bonus for radio astronomers. What was to prove that they could not actually plumb depths completely inaccessible with optical telescopes? After all, if Cygnus A were removed to beyond the range of the 200-inch, it would still be a conspicuous radio object. This was the beginning of the emergence of radio astronomy upon the cosmological scene, that part of the science dealing with the nature and origin of the universe.

Although most astronomers speak of the galaxies as colliding, some Russian physicists prefer the more romantic interpretation that we are seeing a galaxy divide, in much the same way as an amoeba. At all events, there is mutual movement, and what is producing the radiation is not so much stellar collisions (for the stars, even under these desperate conditions, are still relatively far apart) as interaction of the nebulae of the two galaxies. What effect this titanic emission has on the galaxies' planetary populations is rather hard to say.

The identification of Cygnus A was a major triumph for both radio and optical astronomy. Since that time, these two branches of the same science have thrown up the interesting problem of the *quasi-stellar objects*, popularly known as *quasars*, which appear to be extremely remote, super-luminous bodies which show relatively rapid changes in intensity. Astronomers have, as yet, failed to answer most of the questions posed by these astonishing objects. Their red-shifts indicate enormous distances, in many cases greater than the most remote known galaxies; but they do not appear to be associated with any of the well-known galactic clusters. Neither do they appear as anything more than points of light. Although it was once argued that the quasars' red-shifts might be misleading, and that they are really unremarkable bodies quite close to us, further investigation suggests that they may be a form of 'armless' galaxy, all nucleus, perhaps collapsing gravitationally and emitting the colossal amounts of energy that so puzzle physicists.

It is not only colliding galaxies and quasars that emit intense long-wave radiation. The elliptical galaxy M.87 is a fierce transmitter, and there are other examples too. The nearby spiral M.31 is 'normal' (if we may use such a word), and because of its closeness it is particularly easy to study. Results are most interesting. Photographically the greatest extent of the galaxy is about 2°, but radio emission can be traced over at least 6° of sky. A similar extensive radio corona exists around our own galaxy as well, and now that Centaurus A has been analysed we can invoke magnetic fields to explain their existence. They consist of trapped electrons whose movement through the field produces radio waves. It is interesting how we often have to search far afield to explain local effects!

Before *Explorer I* discovered the van Allen zones it was thought that all terrestrial influence ceased with the fading out of the atmosphere at an altitude of about 600 miles. Now it has been enormously extended, and similarly the Galaxy's kingdom is steadily broadening; space is becoming fuller. But it is also becoming larger, and this is a cosmological matter.

CHAPTER 25

The World of Cosmology

THE UNIVERSE is a dangerous place – a sort of abstract wilderness embracing the worlds of physics, astronomy, metaphysics, biology, and theology. These all subscribe to the super-world of cosmology, to which students of these various sciences can contribute. Strictly speaking there is no such person as a 'cosmologist' for the simple reason that nobody can be physicist, astronomer, metaphysicist, biologist, and theologian at the same time.

Nevertheless the essence of modern science is theory based on observation, and in this sense it is the astronomer who manipulates cosmology. It is he who provides the only verifiable evidence of the mechanics of the universe, and his limits mark the limits of the comprehensible universe. The physicist may step in to interpret, to predict, to explain; but without the telescope the physicist is hamstrung through lack of material. Another point to be borne in mind is that the inferences drawn from what the astronomer sees are, in the widest sense, the property of each individual's mind. For instance, an ant's universe is a meagre thing a few yards across. We, with our superior knowledge, know it to be rather bigger; but this knowledge is unintelligible and useless to the ant. Similarly, the human universe may appear absurdly restricted and unimaginative to a mind working on a grander scale. It is a rather disturbing thought that we are, in a sense, trying to equate ourselves with the cosmos, but it would be more disturbing still should we ever succeed.

When we come to consider the birth, or at least the heartbeat, of the universe, we find a crucial contribution being made by the application of radio techniques in investigating very distant galaxies. It has already been explained how radio and visual brightness can be, and usually is, quite independent. The probability is, in fact, that radio telescopes can penetrate to considerably more remote regions of space than can optical ones. Optical telescopes are just not powerful enough to make those crucial observations that may be the prerogative of radio astronomy. But let us take things in their proper order.

Before the beginning of the present century cosmology was a strictly philosophical pastime, and in any case eighteenth- and nineteenth-century thinkers considered the Galaxy to be the entire universe. They were not prepared to believe in the island-universes discovered by Herschel and subsequently interpreted as more homely features. Cosmology was simply a study of the birth and death of the stars, and more important still the birth and death of the solar system (which was the most important product of the universe).

But the sensational advances of the present century have entirely transformed the scene. The discoveries by Hubble and others proved that the essential constituents of the universe are not stars, but galaxies of stars – and perhaps we should now modify this to groups of galaxies. The relativity theory of Einstein showed in essence how the 'empty' space separating the galaxies must be considered as a positive medium. And the application of the spectroscope has shown that the galaxies are without exception travelling away from each other, like fragments of a bomb after the explosion. This, the red-shift of the galaxies, is the very cornerstone of modern theories, and in a sense it has aligned our views of the whole problem.

The shift was discovered by Slipher, at the Lowell Observatory, in 1912: he found velocities of recession of up to 500 miles per second. This, of course, was very puzzling, since in those days the dim spiral objects were thought to be Galactic features, and such a speed was immensely higher than the casual drifting of the stars. In fact it was the first hint that spirals are not connected with the Galaxy at all, a trend of thought to be later proved by Hubble. Subsequent work on fainter and therefore more distant objects rapidly increased the 'shift record', and we now know that all galactic clusters are moving away from each other, the speed of recession being higher the greater the distance. At 30,000,000 light-years, the velocity is 750 miles per second; at 800,000,000 light-years it has increased to 25,000 miles per second. Yet even this is small compared with one of the remotest clusters of galaxies. It lies in Boötes, and seems to be no less than 5,000,000,000 light-years away; its red-shift indicates a velocity of 86,000 miles per second. Every 18 minutes its distance increases by an amount equal to the gap separating the Earth and the Sun!

The farther the faster, is the simple rule of the red-shift, and it is so universal that the method of determining the distance of a remote galaxy is by measuring its spectral shift. From our earthbound viewpoint the galaxies seem to have taken a particular dislike to our system, but this is a very egocentric view, for our galaxy is sharing in the general turbulence as well. The most convenient way of understanding this is to picture a spotted sheet of rubber being stretched. As the surface enlarges so the spots move away from each other, but to any particular spot it would seem as though it alone were being deserted. It is these red-shift observations which have led to the present concept of an expanding universe.

Before going any further it is necessary to ask whether so grandiose a conception can be based on such apparently fragile evidence. Could there not, for instance, be some unknown matter in space which affects the light beams travelling through it, somehow lengthening the wavelength and so producing a shift? In this way we should not have to invoke these colossal displacements of distant sources. Until recently we could only argue that there were no physical grounds for believing in such an effect. But in the past few years radio astronomy has once again come to the rescue. Observations of Cygnus A and the Coma cluster, both of which emit 21-centimetre radiation due to hydrogen gas, show the wavelength to be changed by precisely the amount indicated by visual Doppler-shift measurements. This is excellent independent confirmation, for what affects light waves does not usually interfere with radio waves, and vice versa. We seem to have here an absolute effect that can come only from the sources' motion. There is every hope that continued observation will show radio-frequency shifts for other galaxies as well, and the general consensus of opinion overwhelmingly supports the idea of galactic expansion.

However there is another argument, a rather picturesque one, which depends on the simplest of questions: Why is it dark at night?

As far as our greatest telescopes can penetrate, which is to a distance of several thousand million light-years, galaxies occur with almost monotonous regularity; in terms of number per given volume the visible universe appears to be uniformly populated. Moreover, there is a great deal more beyond. Our view, because of telescopic limitations and other factors (such as the auroral illumination of the night sky), limits us to detection of galaxies within a certain sphere (Fig. 53). If, instead of being at A, our galaxy were at B, we should see a different part of the universe. But there is no reason to suppose that it would

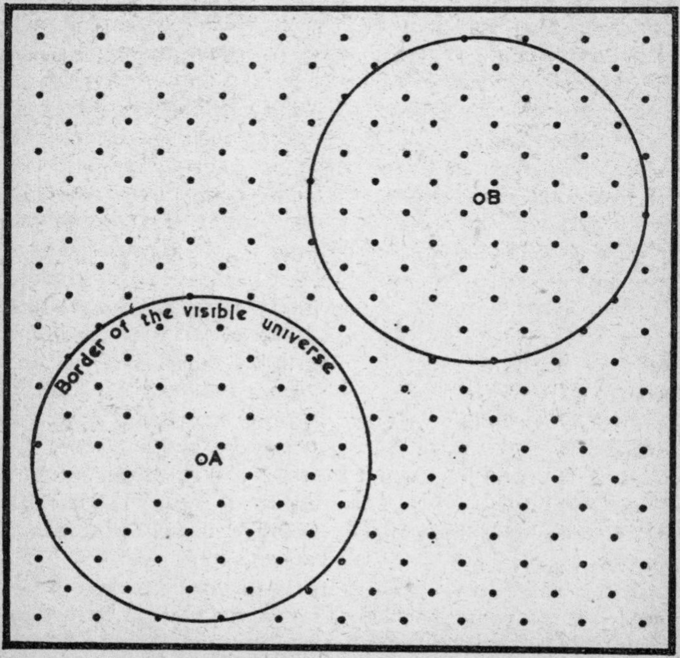

FIG. 53. *Different views of the universe.*

appear fundamentally different, and every reason to be sure that its general features, such as galactic distribution, would be identical.

If we now reject the red-shift observations and suppose the galaxies to be at relative rest, or nearly so, what should be the effect? Those galaxies far beyond our observable horizon will more than compensate for their individual dimness by their numbers; they will mass so densely that their combined light will make the background of the sky glow as brilliantly as the Sun! They will, in fact, form a continuous backcloth to the sky.[1]

One way out of this dilemma is to suppose that their distribution suddenly and spectacularly thins out just beyond the borders of our observable universe. Clearly this is a very clumsy artifice – Ptolemaic even, since it suggests that our own system is the centre of the universe. We must therefore reject it. Another argument is more subtle. Suppose that about 6,000,000,000 years ago, all the galaxies came into being in their present positions. Then the light travelling from objects at distances greater than 6,000,000,000 light-years would not yet have reached us; they would be invisible, and there would not be this cumulative effect of illumination. If we waited for another 1,000,000,000 years we would see 7,000,000,000 light-years into space, so that, in effect, the universe expands before our eyes as light from more and more remote regions reaches us!

But there are many objections to this idea. Why should galaxies suddenly come into being in a universe that already existed? What was in the universe before they were magically formed? What, to put a more metaphysical question, was the point of the universe existing if it did not contain anything? Also it still has not explained the agency causing the red-shift, and all in all the inescapable conclusion is that it is due to motion.

In this case, what happens to the dark-night paradox? It resolves itself very nicely. For if a source of light is moving away from the observer at a velocity comparable with that of

[1] This is known as Olbers' paradox.

light, it follows that the light reaching the observer is appreciably diluted, so that the object appears fainter than it would were it stationary. We might draw an analogy with a leaking can filled with water and dripping regularly. As it is raised above the ground the drips strike the surface at longer and longer intervals simply because each one has farther to fall.

This partly explains why the limit of the 200-inch telescope is fixed at about 5,000,000,000 light-years (and this only in exceptional cases), for at this distance, due to its great speed, a galaxy appears only ¼ as bright as it would were it stationary. And if we assume that velocity continues to increase with distance, an amazing state of affairs presents itself when we delve 13,000,000,000 light-years into space. A galaxy at this distance is receding at the velocity of light, and therefore its light cannot get a start – it is being dragged along behind! Therefore it and all its fellows are invisible, and we do not have to consider an infinite depth of galaxies. But it also has a more significant implication, for the universe, effectively, has a radius of just 13,000,000,000 light-years. We can never know what is beyond this light barrier; it is literally removed from our dimension. We enter the mysterious time–space world of Einstein, which can be explained mathematically but has very little practical meaning.

This final restriction of observation – even though it is to a limit well beyond the capabilities of our present instruments – is a faintly frightening confinement. Our supply of material is exhaustible. It may amount to a million million million galaxies, but it still provokes a nagging worry at the back of the astronomer's mind. Is it a sufficient sample to promote some satisfactory system of cosmology? If not, then solution of this most basic of all problems is forever beyond us, since there is no way of pushing our horizon beyond the concrete limits set by the ever-increasing flight of the galaxies. Moreover, earth-bound astronomers have arrived at the stage where galaxies are being photographed down to the atmospheric limit. The eternal auroral glow douses still dimmer objects, and just as in a thick fog powerful binoculars are no more effective than the naked eye, so it is useless simply building larger instruments. We are looking through a screen of light. Not until a telescope escapes from this veil and is set up in the clear blackness of space can we probe farther, perhaps to 8,000,000,000 light-years. Beyond that the dimming due to recession has so drastic an effect that it would take an unbuildable telescope to drag the galaxies' faint beams from obscurity.

Why should there be this concern, and why do we need to observe such distant objects? Certainly the nearby systems furnish all we need in the way of variety for investigations into galactic structure, and in any case these very remote galaxies appear only as almost star-like points of light. But what astronomers are doing, or trying to do, is to work out their distribution in space, and, ultimately, their distribution in time. For when we look into space we look back in time.

This concept, so vital to the astronomer, is equally unfamiliar to the non-astronomer simply because terrestrial distances are so small. For instance, someone standing in Parliament Square and looking up at Big Ben does not see the time as it *is*, but as it *was* when the light rays left the dial. The fact that this delay is so small – about a millionth of a second – does not alter the fact that it is there. If someone on Hampstead Heath adjusted his watch by looking at Big Ben through a telescope he would be 1/30,000th of a second slow; and once again, since it would take a pretty good watch to notice this error, it can be neglected. Light travels so fast (186,000 miles per second) that the delay does not become appreciable until we start dealing with astronomical distances. For instance, when we observe a sunset we are seeing the Sun as it was 8½ minutes ago, since its light took this time to cover the 93,000,000-mile gap. It has really set even though it is still to be seen on the western horizon.

When we look at the Galactic stars our information is hundreds or even thousands of years out of date, but when we turn to galaxies the position is drastic indeed. M.31 is 2,000,000 light-years away, so that we see it as it was 2,000,000 years ago – long before recognizable men appeared on the Earth. The farther we look, the more we are carried back in time. Our information on the Coma galactic cluster is 100,000,000 years out of date.

This cosmic lethargy of the light beam gives the astronomer an exciting and at first sight almost unbelievable tool: he can actually travel back in time and see how regions of the universe looked hundreds or even thousands of millions of years ago. In probing distance he is also probing time. So he can formulate a theory of the universe, decide how the galaxies should have looked so many aeons ago, and then see how they actually did look.

Like all the best detectives, astronomers have just one great clue on which to base their investigation: the expansion of the galaxies. From this the two main theories, evolutionary and steady-state, have emerged. Of these the evolutionary theory is considerably older, dating from the time when Hubble, in his work on galactic red-shifts, noticed a very curious and suggestive fact: that the speed of recession is directly proportional to distance. We might sum up the galaxies' flight by saying 'twice as far – twice as fast'.

What happens if we run a race, and one of the contestants runs exactly twice as fast as the other? Evidently the faster runner will always be twice as far from the starting post as his rival. And this reasoning led to the 'primeval atom' theory of the Abbé Lemaître, in which he traced the galactic flights back through time and arrived at an epoch, some 10,000,000,000 years in the past, when they must have all been collected together in one region of what we now call the universe. Now it is very important, but exceedingly difficult, to realize that when all the matter was collected into a smaller space (Lemaître suggested it was about 1,000,000,000 light-years across) its dimensions were no more finite than they are today. We can easily see a thousand million light-years into space at the present time, but this does not mean that we could have seen to the 'edge' of Lemaître's first-stage universe! Had we been inside it, it would have still seemed infinite.

Of course this seems rather paradoxical, to say the least of it – and so it is in everyday terms, since we are not used to dealing with such enormous quantities of space and time. It lands us in a world of doublethink, where the part is as great as the whole. But Einstein's theory of relativity solves the problem. In essence, the explanation of the paradox is that space, far from being a negative emptiness, has a very positive restriction on the movements of bodies in it. For instance, an underground train running through a tunnel is forced to follow the tunnel's curves, but by reference to the walls of the tunnel a passenger cannot tell whether he is travelling in a straight line or round a bend. Similarly we can speak of space being 'curved' in such a way that when we recede to great distances it forces us back, or at least slows down our progress, even though we do not realize it. Therefore someone trying to find the edge of the 'small' universe finds it just as impossible as when he is in the 'large' universe. If we imagine a beetle solemnly walking round a circular track, it does not matter if the diameter is large or small; the beetle never finds the end.

So we are back at a time when the material that now forms the galaxies was confined within a much smaller space than it is today. But this was not the beginning of the process; it was merely the end of the first stage. The matter itself came from a far more condensed aggregation that is known as the primeval atom. The primeval atom contained all the material from which the universe is built, so densely packed that its density was perhaps 1,000,000 times that of a white dwarf! It was solid nuclear matter, and once again it must be considered as infinitely large. Worse still, it did not really exist at all, because time did not begin until it exploded and began to form the primordial clouds of hydrogen. If God ever said 'Let there be Light' it was at the instant of the explosion of the primeval atom. What happened before then is beyond the scope of both science and comprehension, since we cannot conceive existence without time; the word itself implies the passage of time.

Why can we not consider the whole process at one swoop? The reason is interesting and very important. Terrestrial experience teaches us that masses attract each other with what is called a gravitational force, which weakens with distance. But there is also, astonishing though it may seem, a repulsive force as well which *increases* with distance. On the terrestrial or even interplanetary scale such an effect is negligible, but on the cosmic stage it comes into play, and is once again a result of relativity theory.

This explains Lemaître's two-stage process. After the initial explosion of the primeval atom, the expansion of the matter was under the control of gravitational forces. This gradually slowed down the expansion until after many thousands of millions of years it effectively ceased. The primordial cloud now started to condense into clusters of galaxies, a process taking an unknown time. As soon as these well-defined masses accumulated, the effect of cosmic repulsion began to make itself felt and pushed the galaxies apart; the forces were now not slowing it down but speeding it up. It is what we see now as the expanding universe.

In very broad outline this is the evolutionary theory that many astonomers are prepared to accept today. Of course, the application of rigid physical principles to such an overwhelming scheme of things is bound to feed objectors. But if we feel a reluctance to discuss events so far removed in both time and scale in such apparently precise terms, the scientist can only answer that it is his task to supply theories to fit observed facts. At our present state of knowledge the idea of a primeval atom, no matter how unfamiliar, succeeds in leading to a theory that does explain observation – even if we must admit that some super-force was responsible for setting the machine in motion. To get over this objection and completely rationalize the universe, a much more recent theory was advanced by British astronomers, among them Fred Hoyle and Hermann Bondi. Hoyle rejects the idea of the universe having a finite beginning – and indeed it seems curious that a process which has to be given an initial push should be capable of proceeding for ever; surely infinity extends in both directions? Instead, he proposes that the universe has been in existence for all time and will remain in existence for all time.

Now there are obvious drawbacks to supposing that the galaxies we see receding into the distance are infinitely old. First, we have excellent evidence that galactic ages are of the order of thousands of millions of years. Second, their infinitely-prolonged motion would have carried them infinitely far away! To combat the fact that galaxies must die, and that their expansion would leave the universe an infinity of emptiness, Hoyle makes a revolutionary suggestion. Matter is created from nothing. As galaxies spread and leave gaps, so new hydrogen atoms are formed to give birth to new galaxies in their stead. They share in the general movement because what is expanding are not the galaxies in their own right, but the space containing them; whatever exists in that space must expand with it. Hoyle's excellent analogy is a doughy pudding filled with raisins; when placed in the oven the dough expands and spreads the raisins. The raisins are galaxies and the dough is space.

The obvious objection to the continuous-creation theory is that physics tells us that matter can be neither created nor destroyed. Actually this is no longer true; mass and energy are mutually convertible. When a hydrogen bomb explodes mass is being destroyed in exchange for energy. But Hoyle's theory demands that matter be created from nothing, not even energy. Where there was previously empty space, there is now a cloud of hydrogen gas. And nobody has yet seen a hydrogen atom form before his eyes – or if he has, he has not yet admitted it![1]

Does this mean the death of continuous creation? Before deciding, we should ask what rate of creation of matter is desired, and the answer can be put in a picturesque way. Suppose we decide to test the theory and install a scientist in the Empire State Building. His job is to take samples of the air in the building to try to catch a hydrogen atom in the process of formation (remembering that each cubic inch of air contains altogether 30,000,000,000,000,000,000,000 atoms of various sorts). Since we could expect the total volume of air in the building to yield only one atom per century, our scientist would clearly have to be rather patient before he could dismiss the idea of continuous creation of matter! Yet this rate is all that is needed to maintain the universe at its present standard of

[1] It is most likely to be a hydrogen atom, since hydrogen is by far the commonest element in the universe, as well as the simplest.

living. Obviously, the steady-state theory could not be disproved along these lines; but equally, could it be proved?

So we emerge on the present scene of the cosmical controversy, which is effectively one of rivalry between the evolutionary theory and the steady-state or continuous-creation theory. It is called steady-state because the crux of Hoyle's universe is that it is not evolving. It is timeless. If we shifted thousands of millions of years into the past, galactic distribution would be the same as it is today – different galaxies, perhaps, but the same general layout. And the same for the future. But if we cast back (say) 8,000,000,000 years in the evolving universe, what do we find? The galaxies have only recently begun expanding from their temporary resting place, and therefore they are closer together. As the universe ages they fly farther and farther apart. In other words, the density of the evolving universe is decreasing, whereas that of the steady-state universe remains the same.

It is here that the astronomer's ability to travel in time becomes of priceless importance, for in surveying regions at a distance of 2,000,000,000 light-years the 200-inch telescope is, according to the evolutionary theory, looking at galaxies as they were when the universe was only $\frac{4}{5}$ of its present age; and the farther we look, the younger becomes the observable universe. This is our key, and this is what astronomers are trying to do: to study galactic distribution at very great distances in space and time and from these studies to decide whether or not the universe is evolving. If in these remote regions galaxies are packed more tightly than in our immediate vicinity, the evolutionary theory is satisfied; if, on the other hand, their density is the same, we must invoke the steady-state theory, whose insistence on constant distribution is sometimes called the 'perfect cosmological principle'.

It is therefore frustrating to find that at just the distance where galaxies might be expected to show a detectable difference (at about 6,000,000,000 light-years) our greatest telescopes are baulked by the auroral glow. The farthest probe is 5,000,000,000 light-years, but this is for an abnormally luminous object; the normal maximum range is about half this.

The point is that if we are to survey the galaxies' distribution properly we must have a fair sample. Our range of adequate visual sampling is simply not great enough.

Therefore it was to radio astronomy that scientists turned in their search for clues. We have already seen that while galaxies are in general of about the same luminosity, some, particularly the quasars, are exceptionally intense emitters of radio waves. Radio astronomers have charted many of these sources, not all of which can be identified with visual objects, and the suggestion is that they are very remote galaxies. The red-shifts shown by the quasars are far greater than those exhibited by any known galaxies of normal type; they are so large, in fact, that Hubble's law relating the red-shift with distance breaks down, and different 'models' of the universe give different distances. It should also be added that some astronomers, unhappy about the vast distances inferred by the quasars' red-shifts, have been questioning whether they are, in fact, due to motion at all. Other theories have been proposed, but none of them have stood the test of observation. At the moment, it is logical to suppose that when we observe quasars and other quasi-stellar objects showing these large shifts, we are probing very deeply indeed into the universe.

The result of all this work has, during the past five years, suggested that the quasars increase in numbers more rapidly with increasing distance than could be expected if the universe were static. It must, however, be pointed out at once that this conclusion is by no means unanimous; quite apart from doubts about the distances of these objects, analysis of their distribution is exceedingly difficult. But another factor has emerged into the scene. Radio astronomers have noted a background 'hiss' pervading the universe, and it is generally accepted that this hiss is thermal in origin; in other words it is purely heat radiation, and is not produced by any atomic processes. If we suppose that the universe began as a primordial atom which exploded, a great deal of thermal energy must have been liberated. Between then and now the density of the universe decreased by a factor of billions of billions, and the radiation temperature fell rapidly too, from perhaps a million

million °C to the −270°C indicated by these radio emissions. There is no way of accounting for this general heat except a 'big bang'.

Something important happened 13 thousand million years ago. We, who are ourselves made up of the same atoms that constitute the stars and worlds of space, are blessed with imaginations capable of asking 'What?' Perhaps, on completing our tour of the visible universe, with its vast time-scale and incomprehensible distances and the majesty of its processes, we find nothing more remarkable than the ability of our puny selves to ponder such questions.

PART THREE

Amateur Astronomy

Although it is true that much of astronomy is the province of the professional worker, using giant telescopes, there is nevertheless scope for useful research by the modestly-equipped amateur. Moreover, it is not necessary to do original work to derive enjoyment from this most vast of all sciences.

CHAPTER 26

Celestial Positions

Before outlining the various spheres of astronomical observation that are open to anyone possessing a pair of binoculars or, preferably, a small telescope, it is necessary to pause a moment and to consider the Sun, Moon, planets, and stars not as they are, but as we see them. For example, few people have much difficulty in explaining the phases of the Moon, but they protest at accounting for why we can see the constellation Orion in winter but not in summer – despite the fact that the reason is probably simpler!

When we look up at the night sky we are not aware that the celestial objects are at different distances. It looks exactly as though they are fastened to a vast sphere, half of which is below the horizon. We can make use of this planetarium effect by inventing a mythical 'celestial sphere' to carry the Sun, Moon, planets, and stars, with the Earth at its centre. The Earth revolves once on its axis in 23 hours 56 minutes, but all motion is relative – so instead we consider the sphere to be revolving, carrying its attached bodies with it. It is therefore evident that a star will return to the same position in the sky after an interval of 23 hours 56 minutes, and this is called the sidereal day ('star day'). It is divided into 24 sidereal hours, which are slightly shorter than ordinary civil hours.

Just as geographers have marked imaginary lines of longitude and latitude on the Earth's surface, so astronomers have divided up the celestial sphere. By prolonging the Earth's axis to meet the sphere, the north and south celestial poles are marked; the celestial equator is inscribed by extending the plane of the terrestrial equator. Lines of celestial latitude or 'declination' (Dec.) follow, running up to 90° N and S. On the

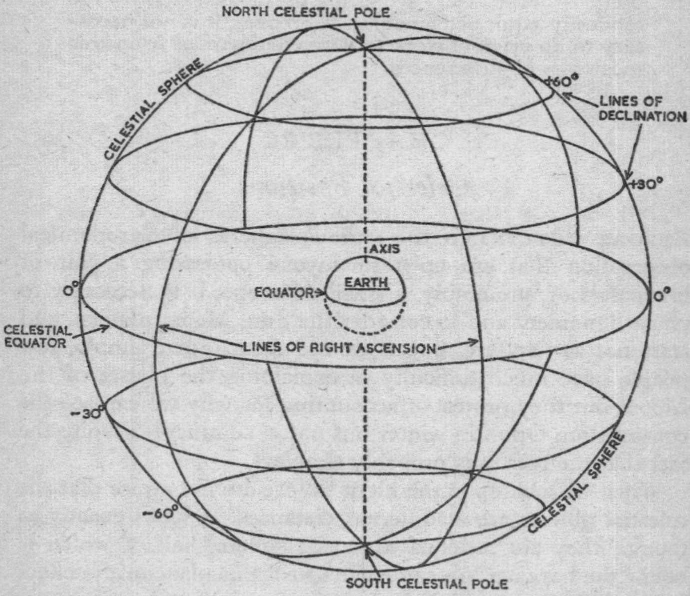

FIG. 54. *The celestial sphere.* To avoid confusing the diagram still further, only a few lines of RA and Dec. are shown.

Earth longitude is reckoned in degrees also, but the celestial sphere has 24 fundamental intervals of 'right ascension' (RA), to signify the 24 sidereal hours of its rotation.

The sphere, thus compiled, is shown in Fig. 54. However, it still lacks one other basic line which marks the Sun's annual path. This line is known as the 'ecliptic'.

Since the Earth revolves around the Sun, the effect is to

make the Sun appear to circle the celestial sphere. At first sight it may seem that the ecliptic should coincide with the celestial equator, but this is not so because the Earth's axis is tilted, at an angle of $23\frac{1}{2}°$. This produces the seasons (Fig. 55). When the north pole is at its maximum presentation to the Sun (in June), it is northern midsummer; 6 months later it is turned away, and the northern hemisphere experiences winter, while the southern hemisphere has its summer. Therefore the Sun spends half the year north of the celestial equator, and the other half south, so that the ecliptic is tilted with respect to the

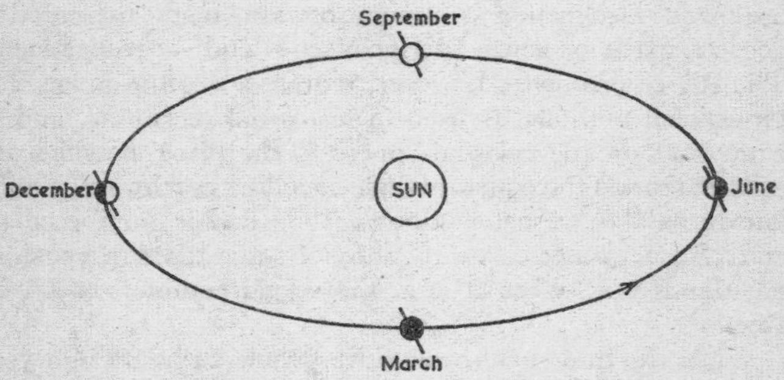

FIG. 55. *The seasons.* For purposes of clarity the Earth's axial tilt is somewhat exaggerated.

equator. The angle, $23\frac{1}{2}°$, is naturally enough the same as the Earth's axial tilt.

The ecliptic is really nothing more than a reflection, in the sky, of the Earth's orbital plane; it defines the level of the solar system, at least approximately, while the celestial poles and equator are produced by the Earth's own idiosyncrasy in having an axial tilt at all. It would be much simpler were the axis upright, since ecliptic and equator would then coincide; the drawback would be the lack of seasons! At any rate, since most of the planets have orbits lying in almost exactly the same plane as the Earth's, we should expect them, in their slow crawling across the celestial sphere, to keep to the ecliptic. This they do, to within a few degrees, and by drawing a belt some 18° wide around the ecliptic (the Zodiac), we can ensure that they are always found within its confines. Unfortunately the eccentric Pluto, with its large orbital tilt, can escape, and so can some minor planets, but on the whole the Zodiac retains its planetary population well.

The inferior planets present a special case, and remain dependent on the Sun, but since we are inside the orbits of the superior planets they circle the Zodiac in their respective years: Jupiter takes $11\frac{1}{2}$ terrestrial years, while Pluto requires $2\frac{1}{2}$ centuries.

There remains the problem of dividing the sky up into co-ordinates. Declination is straightforward, being measured in degrees north or south (designated + and − respectively). The RA co-ordinates, however, require a starting point. The Greenwich meridian is used in terrestrial reckoning, and its equivalent on the celestial sphere is the place at which the ecliptic crosses the equator when travelling northward, a point known as the 'vernal equinox'. This line is numbered 0^h, and RA is reckoned eastward. It follows that the Sun's position on March 21st, when it is at the vernal equinox, is RA 0^h, Dec. 0°.

While the Sun runs through the whole 24 hours in a year, and the planets in their periodic times, the stars remain virtually unchanged. We must say 'virtually' because there is actually a slight drift of the whole sky due to a progressive shift of the Earth's axis known as 'precession', but for most purposes this is negligible. Moreover, stellar proper motions affect their co-ordinates. However, the position of Sirius has only changed from RA 6^h 41^m, Dec. −16° 35′ in 1900 to RA 6^h 43^m, Dec. −16° 39′ in 1950. Most star positions are given for the epoch 1950, because the change is so small that they are still good enough for most purposes.

The fact that we see different constellations at different seasons of the year follows directly from the Sun's circuit of the celestial sphere. For we can only see stars in that part of the sky away from the Sun, and as it moves eastward across the sky it progressively drowns some constellations and reveals others. In June, for instance, it is in that part of the ecliptic that passes through Taurus, and for a couple of months on either side of that date the constellation is placed in the daylight sky. By December, when the Sun has moved round to the opposite part of the heavens, the Bull glows brightly in the southern sky at midnight.

This shift can be thought of as occurring through the

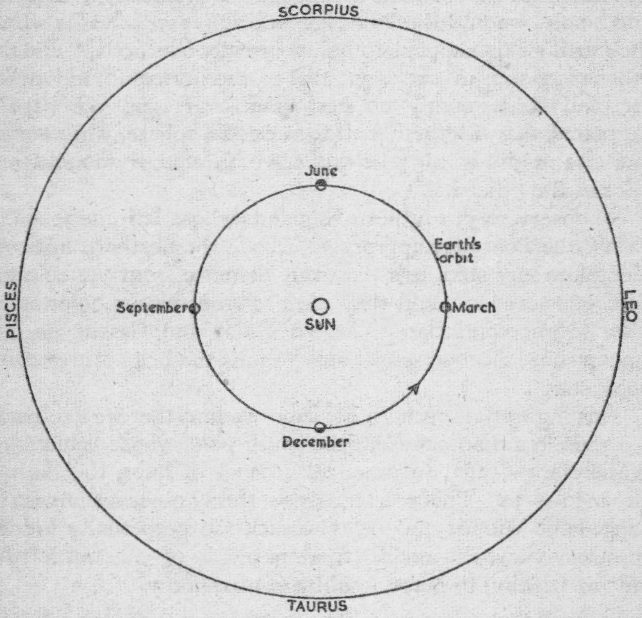

FIG. 56. *Seasonal drift of the constellations.* The four asterisms shown here are simply picked at random.

difference between sidereal and solar time. The Sun is due south every day at roughly 12 noon, but a star reaches the same position 4 minutes earlier each day (thereby explaining why the sidereal day is only 23 hours 56 minutes long). Consequently, since we measure our days relative to the Sun, the constellations slowly progress from east to west. After a year they have advanced so far that they are back where they started from. There are therefore 365 solar days in a year, but 366 sidereal days.

Finally there is the matter of 'circumpolar' stars. An observer on the equator will theoretically see both celestial poles at his north and south horizons, and every star in the sky will rise above the horizon during the sidereal day – although many will of course be drowned by daylight. If we now travel to the north pole, we find the Pole Star (a bright star towards which the Earth's axis happens to point) directly overhead, while the other stars spin in circles parallel to the horizon. They never set, and consequently no new stars ever rise. We have a perpetual view of exactly half the celestial sphere, while someone else at the south pole will have an equally monotonous view of the other half.

To observers in southern England, whose latitude is about 52° N, the Pole Star appears 52° above the northern horizon. Therefore any stars less than this distance from the celestial pole can never set, and they are known as circumpolar stars. Two bright constellations, Ursa Major and Cassiopeia, are among those always visible, and Vega is the brightest circumpolar star.

Turning to the southern horizon, we find that we can never see stars less than 52° from the south pole, whose declination is therefore $-38°$. In practice, because of haze, the limit is closer to $-30°$. This means that southerly objects such as the Magellanic Clouds and the Coalsack are perpetually hidden from our view. But luckily there is plenty of material left for anyone wishing to make a hobby of astronomy.

CHAPTER 27

Naked-Eye Astronomy

ASTRONOMY IS a remarkably accommodating science, and with the possible exception of geology it is the only one in which amateurs and professionals both have their private parts to play. This may come as a surprise when one remembers the vast sums spent on large instruments (the 200-inch telescope, completed in 1948, cost about £2,000,000, and would undoubtedly be even more expensive today), but it is less so when the vastness of the field is considered. If modern telescopes have pushed back its frontiers to the extent of thousands of millions of light-years, they have necessarily had to gloss over some of the nearer regions in the process. It is hard to say which might be the more important: the detection of a yet more remote galaxy, or the early observation of a bright nova. But we can say that an amateur observer has a good chance of discovering the latter and handing it over to professional scrutiny before an observatory patrol camera catches it. This is not mere brazen competition; even a few hours' priority may make all the difference to our benefit from that particular event, for a nova has never yet been caught really early in its drastic rise to splendour.

At the same time it would be futile to pretend that the amateur of 1971 has as wide an opening for serious research as he would have had in 1900. The observation of double stars; much planetary and virtually all lunar and solar research; studies of clusters and galaxies – all these fields and many more have been swallowed up by improved techniques. Yet there is still a good deal of work left, especially in the study of meteors and variable stars, to say nothing of comet-sweeping. Anybody who is prepared to buy even a pair of good binoculars, costing perhaps £15, will find enough to keep himself occupied for a lifetime. But before then, if he is wise, he will spend some time getting to know the constellations and their brighter stars, as well as the positions and slow drifting of the nearer planets. There are several excellent books about observational astronomy, but a few remarks here may not be out of place.

The first and greatest essential is a star map, and by far the best is Arthur Norton's *Star Atlas and Reference Handbook*, published by Gall & Inglis. Usually known as *Norton's*, it covers the entire sky in 16 charts that show every star visible with the naked eye (as well as fainter nebulae and star clusters), includes notes on interesting objects, and gives a great deal of information about all departments of astronomy. When getting to know the sky a very useful accessory is a planisphere, a device that can be 'dialled' to show the constellations visible at any particular moment. It is also of value when trying to find a comet or planet that is situated in the twilight sky.

Another important item in the amateur's equipment is a torch whose bulb has been dimmed with green cellophane or some other suitable filter (green is the best, since it is the colour to which the eye is most sensitive).[1] It is very necessary not to use more light than is absolutely essential for making notes and looking at the map, since bright illumination temporarily blinds the eye and makes it insensitive to faint objects. For this reason it is necessary to spend five or ten minutes getting the eyes thoroughly dark-adapted before trying to observe something faint and diffuse such as an aurora, while under first-class conditions, when the sky is really black, it may take up to half an hour for the eye to become fully conditioned. Town dwellers are less fortunate, of course, and will look in vain for the fainter celestial objects; but there are many bright ones, and in this country some of our most prolific observers of planets and variable stars work in an unattractive aura of mercury or sodium street lamps. An astronomical enthusiast worthy of the name will always find a means of overcoming difficulties, whether instrumental or environmental,

[1] Amateur astronomers are notoriously makeshift. The usual method is hastily to wrap the torch inside a convenient handkerchief, with disastrous results when it slips out.

and useful and interesting work can be done with naked eye or telescope, in town and country.

Even though early observations may seem very casual and unimportant, it is an excellent idea to keep some sort of journal right from the beginning. If drawing constellations or star clusters the best answer is a book with adjacent pages ruled and blank, so that notes can be written beside the sketches. A general notebook, in which mention is made of all the objects observed, will be of great value later on when it is decided to concentrate on one particular field of study; old observations can be consulted and compared with more recent ones – often with spectacular results. Every observation should be accompanied by date and time, the latter given in Universal Time (UT), which reaches 24^h at midnight. Never use Summer Time. If binoculars or a telescope are employed there must also be details of aperture and magnification.

'Learning the constellations' may sound a fearsome undertaking, but it is really astonishingly easy to remember the major ones, and once these are known the lesser asterisms quickly drop into place. Different stars become known, either for their brightness or their colour, and as the seasons pass so new parts of the celestial sphere creep into view in the east while others vanish in the west. The heavens are never still; what is more, the yearly rotation of the sky presently brings remembered friends instead of strangers. In a way it is just as exciting to glimpse Castor and Pollux, the Twins, in the September dawn sky as it is to witness a brilliant meteor or an eclipse; and this side of astronomy – its quiet beauty – is something that the amateur can appreciate in full measure.

Quite apart from accomplishing the essential groundwork of finding one's way about the sky, there are some interesting fields of practical observation open to anyone lacking a telescope but possessing plenty of patience. Watching for meteors is one. Since a meteor can appear anywhere in the sky without warning, telescopes are completely useless, and there is still a certain amount for the amateur to do. Photographic and radar techniques have entirely supplanted the visual plotting of trails, but simple counts of the number of meteors seen per hour during a shower is not to be despised, while there is always the chance of an unexpected burst of activity, as with the Phoenicid shower in 1957. Moreover, some showers vary considerably in activity from year to year, and cannot always be predicted with accuracy.

Should a fireball burst into view, as it may do when the observer is engaged on some other observation, it is worth immediately marking its apparent path on the star chart. If other nearby observers have done the same, later co-ordination of the observations may lead to the calculation of its real path through the atmosphere – and even to its discovery. Very bright meteors sometimes leave a train that may take several minutes to disperse, and these are worth watching out for.

The aurora is another semi-atmospheric phenomenon to provide useful work for the amateur if he lives in the country, well away from artificial lights; if he lives in the north of England or Scotland, so much the better. In this case it is well worth while making a routine of glancing at the northern horizon during a night's work and keeping a regular 'aurora book', noting negative sightings as well as displays. It takes only a moment, and can amount to a valuable record – especially if done in conjunction with regular solar work.

There are many classes of aurora, and if an active display occurs note should be taken of its extent at different times, as well as colours, arcs, rays, and other features. Recently, with sunspot activity near maximum, aurorae have been seen quite frequently, but they will become less common towards about 1975. They also provide interesting scope for photography, especially if colour film is used. For a bright display, using a high-speed emulsion, the exposure can be as short as 30 seconds at f/2. The trouble is that most aurorae move appreciably even during this short period, so that the problem is a complex one, and well worth investigating.

Turning to the true depths of space, we find some of the bright variable stars excellent prey for the naked eye. Lists are given in many books, but among the most interesting brighter variables may be cited γ Cassiopeiae; o Ceti, when near maximum; α Herculis; β Leonis (suspected); α Orionis;

β and ε Pegasi, and ρ Persei. The method is to judge the brightness of the variable against nearby stars of approximately the same magnitude, and to estimate its place in the sequence. Suppose we use two 'comparison stars', as they are called: A, magnitude 2·3, and B, magnitude 2·7. If the variable appears exactly midway in brightness it is 2·5; if closer to A than B it is noted as 2·4.

This procedure sounds nice and simple, but in practice there are snags. For instance, if the variable is unusually bright there may be no suitable comparison stars near by; Betelgeuse is a case in point. Another difficulty is colour. Magnitudes are determined photographically, and on this basis a red star and a white star may be given the same magnitude but appear appreciably different to the eye, the red star seeming to be the brighter. Because of this it is obviously wise to select comparison stars of the same colour as the variable, and sometimes this is very difficult. In recording the observation it is essential to state what stars are used, and also the catalogue from which their magnitudes are obtained, since other observers may employ different values.

All the stars listed vary slowly, and observations need not be made more often than once a week; they do not take long to make once the comparison stars are selected, and it is fascinating to watch the light curves gradually begin to fluctuate. To begin with, estimates will probably vary from night to night – clearly nothing to do with the star! – but with practice it is possible to be accurate to within $\frac{1}{5}$ or even $\frac{1}{10}$ of a magnitude, depending on circumstances. Some stars must necessarily be lost for a few months in the year, but others, such as γ Cassiopeiae, are circumpolar and permit uninterrupted observation. None of these variables are studied regularly at professional observatories, and it seems a pity that so few amateurs keep an eye on them.

It is all very well to observe systematically, but observations which remain tucked away in a notebook might as well not have been made at all; they must be either co-ordinated with the work of others, or else published in their own right. The only way of putting them to use is to belong to an astronomical society, and the most important amateur organization in Britain is the British Astronomical Association. The BAA is essentially an active body, and to this end it is divided into different Sections that cater for everything from comets to aurorae. By joining the BAA the newcomer to astronomy will not only gain an outlet for his observations; he will be able to exchange ideas with other observers and so enhance his own standing as an amateur. It is true today more than ever before that amateur astronomy is a pursuit of co-operation. There is little the average lone worker can do better than a group, and by keeping his observations to himself he is simply depriving others of his experience. Therefore, when the time comes to buy a telescope and bring the stars nearer, it is well worth forfeiting a little extra in joining this society of amateurs.

CHAPTER 28

Starting an Observatory

A CENTURY ago, in the heyday of the amateur astronomer, many private observatories were better equipped than professional establishments. Since then the perspective has changed very much, thanks to the growing awareness of governments and wealthy foundations that astronomers are not all wizened, grey-haired specimens who are only to be seen when the Moon is out. Astronomy is very demanding in its choice of instruments, and the great sums that are being spent, especially in the radio department, are good evidence that its future is well catered for.

Correspondingly, amateur observatories have gone into a decline. The traditional domed structure has almost disappeared from the scene, to be replaced either by a garden shed in which the telescope is stored, or, more often, nothing at all. This is much less romantic, but it is more economical – and it leaves more to be spent on the telescope itself, which is, after all, the nucleus of the whole affair.

Buying a telescope may sound a straightforward business, and up to a point it is; but there are many traps for the unwary. A lot of unwise purchases have been made in the past, and will be made again in the future, simply through lack of advice. It is worth taking a little care in selection before signing a cheque for an expensive instrument, and curiously enough it is not always the most costly purchases that are the wisest.

The usual 'refracting' telescope, which is the kind sold also for terrestrial purposes, is shown in Fig. 57. It employs a large lens known as the 'object-glass' to focus the light from the object being viewed, and a much smaller lens, the 'eye-piece', to magnify the image and feed it into the eye. If the telescope is a good one, both these lenses consist in themselves of two or more lenses made of different kinds of glass. This is because a simple lens produces an image surrounded by a halo of false prismatic colours, which is almost entirely banished when two or more lenses are combined. Such compound lenses, corrected for colour effects, are said to be 'achromatic'.

Such a telescope will actually give an upside-down view, which is not very satisfactory for normal purposes, and so an extra lens system is incorporated inside the tube to make everything erect. This is omitted from an astronomical tele-

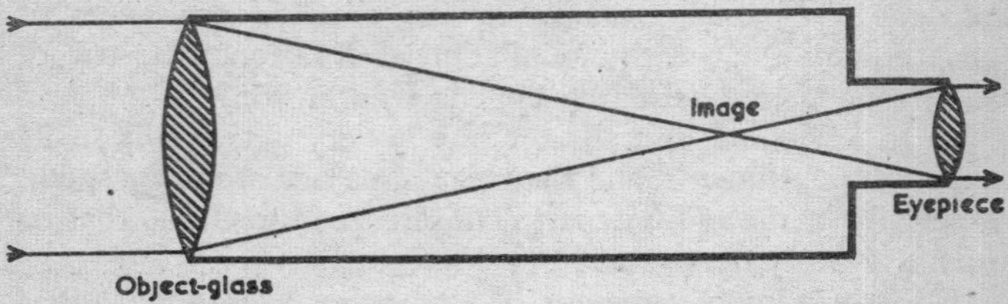

FIG. 57. *A refracting telescope.* For reasons of space the tube length has had to be compressed, while the eyepiece is much smaller than this drawing suggests.

scope, because here light is all-important; we cannot afford to waste the slightest scrap when dealing with so faint an object as a star, and whenever a light beam passes through a lens it is slightly weakened. All astronomical telescopes therefore give an inverted view. At first this takes some acclimatization, but there is, after all, no up and down in space, and the fact that the Moon and planets appear with south at the top is of little consequence.

The other type of instrument, the 'reflecting' telescope, is less familiar to the non-astronomer. Here (Fig. 58) we have, instead of an object-glass, a concave mirror at the bottom end of a tube whose top is left open. This mirror focuses the light by reflecting it back up the tube, where it is met by a small plane mirror called a 'flat' and reflected through a hole in the side of the tube, where the eyepiece is situated. Therefore instead of looking up the tube from the bottom end, the observer peers into its side near the top end. This is a much more comfortable position in which to observe, especially

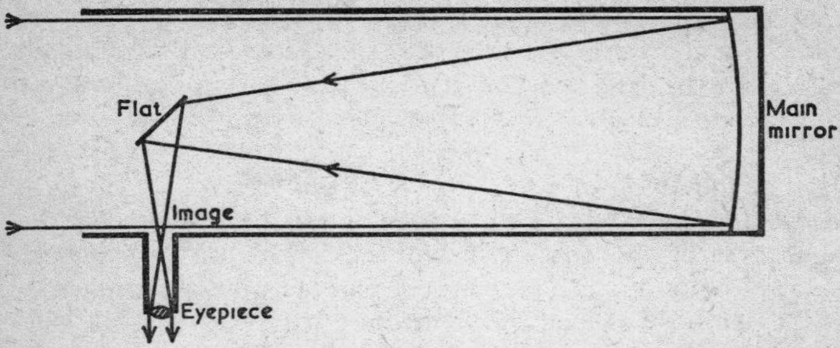

FIG. 58. *A reflecting telescope.* The curve of the mirror is considerably exaggerated.

since there is no agonizing back-bending when squinting at an object almost overhead. In addition the tube of the reflector is usually relatively short, so that it is easier to mount satisfactorily.

Aperture for aperture reflecting telescopes are much cheaper than refractors, and the difference lies mainly in the relative cost of mirror and lens. A 6-inch diameter mirror can be bought new for about £15, while a 6-inch object-glass might cost ten times this amount. Mirrors and lenses are always rated by their aperture, since this affects the amount of light they receive and hence the brightness of the resultant image. Also, because of the nature of light, a large aperture will show more fine lunar and planetary detail than a small one, while it will also resolve (i.e., show as separate stars) closer double stars. For instance, a 2-inch telescope will show that γ Andromedae is double, but it takes a 12-inch to reveal the companion as a close double in itself.

While reflectors are cheaper than refractors of equal aperture, they are also less efficient; for instance, a 3-inch refractor is a powerful instrument, capable of showing much lunar and planetary detail, but a 3-inch reflector is of little use. This comparison is less extreme in larger sizes, but the minimum useful aperture for a reflector is 6 inches. Also, they are more trouble to look after. Both the mirrors are coated on their front surfaces with a reflecting layer of aluminium, and in time this tarnishes and has to be replaced. They also have to be very precisely aligned in the tube. So do the lenses of a refracting telescope, but the reflector tends to come out of adjustment in a few weeks,[1] whereas object-glasses seem to stay stable for years.

The mounting of a telescope is of the greatest importance, since a shaky tube will show nothing satisfactorily except the shortness of the observer's temper. There are two distinct types of mounting. The simplest has the tube swinging vertically in a fork which itself revolves horizontally, both axes being equipped with 'slow motions' which allow the telescope to be shifted very slightly so as to follow the drift of the star or planet. It is always a revelation to the ordinary citizen to see how quickly the Earth rotates! With a moderately high power such as × 300, a star is carried right across the field of view in 30 seconds or less, so that the observer is kept fully occupied operating the slow motions to keep it in view. This soon becomes automatic, but even so it means that he really needs five hands (two to manipulate the controls, one to hold the torch, one to hold the notebook, and one to draw with).

To solve the problem of the so-called 'altazimuth' stand, the 'equatorial' mount was invented. Its principle is simple. The diurnal shift of the stars comes about through the Earth's east-to-west rotation upon its axis. Accordingly, if we align one of the telescope's axes (the polar axis) with that of the Earth, and drive it around this axis so that it revolves in the opposite sense to the Earth, the net result is to keep the telescope stationary with respect to the stars. This drive can be mechanical, using a geared-down gramophone motor or some similar device to rotate the polar axis once in 23 hours 56 minutes, or it can of course be supplied by hand. If a motor is used, it means that the object stays firmly fixed in the field throughout the night.

There are several different patterns of equatorial mount, some suiting refractors better than reflectors, but the principle is in all cases the same. It is clearly a great advance over the

[1] This is by no means always the case, but some reflecting telescopes seem to have a strange spirit of their own.

simple altazimuth, and if celestial photography is being attempted, when the camera must be held absolutely motionless relative to the stars for long periods of time, it is a necessity. On the other hand, it is suitable only for permanent telescopes built on to a concrete pier or something equally solid, for the polar axis must be set with considerable accuracy. It is also possible to over-glorify the equatorial, for many observers have produced incomparable observing records with altazimuth instruments. Herschel is a case in point.

When it comes to actually selecting the instrument, the choice depends to an enormous extent on the type of observing that is anticipated. For a newcomer to the science who wants to find his way about the sky and have a look at everything within sight (which is how almost everybody starts), a small refractor on an altazimuth stand is the ideal instrument. The usual aperture is 3 inches, which means a tube about 3 feet 6 inches long. The 3-inch refractor is by far the most common and universal instrument; it is compact enough to be portable, but sufficiently large to show planetary detail and a great many nebulae, clusters, and double stars. A $3\frac{1}{2}$-inch is of course superior, but when we reach apertures of 4 inches and over the telescope becomes too bulky for a wooden stand, and really demands a permanent, rock-steady mounting, which immediately negatives one of its advantages. After all, few people have a completely clear horizon, and the great virtue of a 3-inch is that it can be moved around to avoid obstacles.

Small refractors, particularly second-hand ones, often come fitted to a dubious erection known as a table stand. This is about a foot high, and consists of a central brass pillar with a universal joint at the top, supported by three collapsible legs. This in turn is supposed to stand on a table, hence the name. This ingenious device is rendered useless by two circumstances. First, the stand or the table (or both) is hopelessly unsteady, so that the stars perform a celestial jive; second, when looking at a high object the eyepiece is so low that the observer has to grovel on his knees. The person who invented the table stand was certainly no astronomer, but manufacturers have been plaguing amateurs with them for two centuries.

The proper mounting for a refractor is a tall, massive tripod, with its legs in one piece; folding legs invariably have slight residual shake, and the smallest movement is magnified a hundred times when using the telescope. Ideally the eyepiece should be at eye level, and this can be arranged by adjusting the angle of the legs. Tripods are invariably made too short, and the legs should be at least 6 feet long, for the slightest stoop assumes agonizing degrees after just a few minutes. Slow motions, which take the form of extendable rods fixed to the eye end of the tube, are very convenient when a high magnification is used, but are apt to be cumbersome when ranging freely across the sky with a low power.

Moving to reflecting telescopes, the rough equivalent of a 3-inch refractor is a 6-inch reflector. This is certainly the smallest useful size, and a warning must be given against buying some of the tiny 3-inch and 4-inch reflectors now being manufactured. These are, emphatically, not astronomical instruments; they are toys, and many are hardly, if at all, superior to binoculars. What is more, they can cost as much as a good second-hand refractor, and the latter will give a lifetime's consistent service.

New 6-inch reflecting telescopes are being manufactured, but they are relatively expensive, costing anything from £50 to £150, while a new 3-inch refractor will cost over £50, if it is of good quality. But generally speaking it is preferable to buy a second-hand instrument, since telescopes depreciate considerably in price but not at all in quality. It is, in fact, almost the opposite. The production of a first-class mirror or object-glass is a task to challenge the most expert optician, and it is a process that cannot be entirely mechanized, so that many lenses produced by veteran craftsmen are even better than those made by their contemporaries. At the very least it can be said that newness is no assurance of optical quality, and a telescope should never be bought without first of all giving it a thorough test. If this is not possible, then go elsewhere. To buy one 'blind' is like purchasing a gramophone record without hearing it.

Testing a telescope is not a difficult process, but it requires a certain amount of experience and is ideally delegated to an astronomical friend (once again an advantage of joining a society such as the BAA). Broadly speaking it consists of examining the image of a star when the eyepiece is placed both inside and outside the position of true focus; the star expands into a disk, and the distribution of light in this disk gives a key to the quality of the object-glass or mirror. A high magnification must be used, and this brings up the question of eyepieces.

The magnification of a telescope varies with the 'focal length' of its eyepiece, and the procedure for determining the power given by any particular eyepiece is as follows. Point the telescope at the Sun with the eyepiece removed, and hold a piece of paper just beyond the drawtube, where the eyepiece would normally be situated. Adjust the position of the paper until the image of the Sun is perfectly sharp, and measure the distance from the paper to the mirror or object-glass. This gives its focal length, which is usually about 40 inches for a 3-inch refractor or 48 inches for a 6-inch reflector. The focal length of each eyepiece is marked on its side, and by dividing this into the telescope's focal length, the magnification is found. For example, if a $\frac{1}{2}$-inch eyepiece is used on a 3-inch of normal focal length, the magnification will be $\times$ 80. If the focal length is only $\frac{1}{4}$ of an inch, it will give a power of $\times$ 160. The important thing to remember is that magnification is independent of the aperture of the telescope; it depends solely on the focal lengths of its mirror or object-glass, and the eyepiece.

Eyepieces are extremely important, since if they are of poor quality the resultant image will be poor, and it may be unjustly blamed on the objective. It is a very curious but common fact that good telescopes are often provided (by their makers) with bad eyepieces that ruin their performance.[1] Care must be taken in selecting only good eyepieces, and they usually cost about £4–£8 each; the Japanese 'Swift' eyepieces are a good type. Some of those supplied with ex-Government telescopes

[1] The writer knows of an excellent professionally-owned 8-inch refractor whose eyepieces are so poor that it is essential to take his own.

are excellent for astronomical work, and are considerably cheaper, but their selection is largely a matter of luck.

It is obvious that different objects require their own special treatment; a whole-disk view of the Moon demands a low magnification and a wide field of view, while Mars must be viewed under a high power if details are to be made out on its tiny face. Ideally, a 3-inch refractor should have four eyepieces, giving magnifications of about × 30, × 80, × 140, and × 200. At first the irresistible temptation is to always use the highest possible power, but disillusionment comes quickly; the atmosphere is never perfectly steady, and on most nights a powerful eyepiece will show Mars as nothing more than a boiling blur of light, whereas a rather lower magnification will show the details considerably better. If only three eyepieces can be afforded it would be best to sacrifice the highest power, while if a comet comes along a very low magnification, × 15 or × 20, will be much appreciated.

A good second-hand 3-inch refractor, mounted on a firm wooden tripod with an altazimuth head, should not cost more than £30, and may possibly be even less; bargains are sometimes to be found in publications such as the *Exchange & Mart*. Professionally-made reflectors are harder to come by, since there is a strong tradition in home construction, and prices vary greatly. Of course, anyone who is sufficiently determined can buy a mirror and flat and make the rest himself.

All telescopes should be fitted with a 'finder', which is simply a small, low-power telescope fitted to the main tube near the eyepiece, so that the main instrument can be directed quickly at the object; anyone who has tried to find a star using a high-power eyepiece will passionately testify to its necessity. The finder should be adjustable, so that it can be aligned accurately, and it should not be too small; for most purposes an aperture of at least 1½ inches is desirable, but one rarely finds makers agreeing on this point, and it is often advisable to fit a larger one.

Refractors are hardy instruments, and can be badly ill-treated without coming to any harm, but reflectors are more delicate. Both the main mirror and the flat must be covered when not in use, and if the telescope is stored out of doors, either in the garden shed or beneath a tarpaulin, it is a good idea to close up the tube completely by screwing an old eyepiece into the drawtube and (in the case of a reflector), putting a lid over the open end of the tube. Astronomical telescopes have a peculiar fascination for insects, and if the tube is left open one sooner or later has an astonishing view of an enormous earwig gyrating against the sky, which has to be emptied out of the eyepiece. Eyepieces, incidentally, should be stored in a special box where they cannot roll around and damage each other.

If a permanent concrete-based telescope is being set up, its position must depend on the type of observation that is going to be carried out. Lunar and planetary work demands a clear south view, since greatest altitude is reached in this direction, while observation of Venus requires good east and west views as well; the north is unimportant. However, variable-star observers need the north rather than the south. In most cases it is a question of compromise, since few amateurs have much choice in the way of horizon obstruction and must simply make the best of things. There are also more general influences; for instance, no town-based observer need consider the observation of faint variable stars or comets. On the other hand, the smoky pall over built-up areas often produces a very steady atmosphere, and is conducive to lunar and planetary work. Every site has its disadvantages, but no amateur worth the name will let himself be deterred by drawbacks.

CHAPTER 29

Amateur Astronomy – the Solar System

ONCE EQUIPPED with a telescope and a working knowledge of the heavens, the amateur astronomer finds several fields of work awaiting him in the night sky. However, there is no real need to await nightfall, for the Sun is a source of constant interest. The user of a 3-inch refractor can find plenty of activity in its developing and decaying spots, and there is always something new to see. Direct observation is of course out of the question, and there are two methods: projecting an image of the Sun on to a sheet of white card, or observing through a device known as a solar diagonal, which removes most of the light and heat from the image. Of these the projection method is normally the best, because it allows the positions of all the spots to be drawn with considerable accuracy.

A 6-inch image is probably the most convenient size, and using a medium-power eyepiece (which must, of course, include the entire Sun in the field), the card will need to be held about a foot behind the drawtube. The best way is to make a light wire frame to hold it in position, not forgetting to place another card higher up the tube to shield the image from the direct solar rays. A circle of standard size is drawn on the card, crossed by a number of fine lines to act as a position grid, and the sunspot positions are copied on to a similar circle and grid drawn in the observing book. It is here that the advantage of a clock-driven equatorial mounting becomes apparent, for the Earth's rotation inches the Sun across the card and the telescope must be continually adjusted.

Once they are accurately located the features of the spots can be drawn in, taking particular note of the outlying pores – since it is here that drastic changes often occur. Very tiny spots that would otherwise be overlooked can often be brought into visibility by slowly swinging the image from side to side, the movement attracting the eye. It is also necessary to orientate the drawing, and this is done by letting the image drift across the card and marking the preceding (west) and following (east) points on the limb. These do not, however, mark the solar equator, for the Earth's axial tilt means that its inclination changes throughout the year. Fig. 59 shows the way its position alters.

When daily observations are continued for several weeks the larger spots come to be expected at their reappearances at the eastern limb. Often they have changed their appearance so completely that they are unrecognizable except by position,

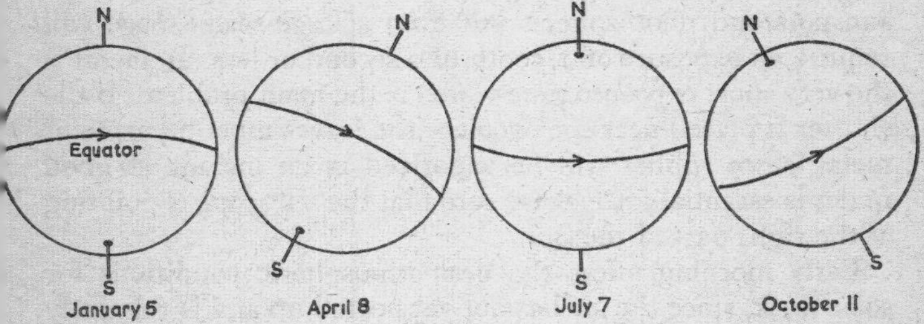

FIG. 59. *Different views of the Sun.* This shows the naked-eye or binocular view, with the Sun due south. The spot paths are straight on about June 5th and December 7th, the date varying due to leap-year adjustments.

and they may of course have disappeared altogether. When a spot is very near the limb it is worth making a special study of it to see if it shows the Wilson Effect. Very little is known about this phenomenon (some spots show it, others do not), and a detailed study extending over a long period could be of tremendous value. Also, a large spot approaching the central meridian is the sign to look out for aurorae.

A solar diagonal is essential if individual spots are to be studied in detail; it consists, essentially, of an unsilvered glass 'mirror' which reflects only 5 per cent of the light into the eyepiece. Many makers sell what are called Sun caps, which are thick disks of dyed glass that fit over the eyepiece, but these are extremely dangerous, as well as being almost useless from a practical point of view. If used for prolonged periods they become extremely hot, and the thick glass expands and tends to snap – with obvious results. Moreover, they interfere with the field of view, and if regular close-up work is envisaged the essential answer is to buy a proper diagonal. Unfortunately, they are rather expensive.

The Sun offers clear scope for photography, and some resourceful amateurs have taken excellent pictures with very primitive equipment. An old plate camera, minus lens, is ideal, since the solar image can be focused directly on the screen. Very slow orthochromatic plates are best, since they are sensitive to yellow light and improve the contrast between sunspots and photosphere, but even a large-scale image will require an exposure of 1/500th of a second or less. In fact it is the very short exposure time which is the main problem. If the shutter is placed near the eyepiece the leaves must be made of metal, since rubber will be vaporized in an instant. A good finder is essential for making sure that the telescope is pointing at the right part of the Sun.

Early morning offers the best atmospheric conditions for solar work, since the air has not yet heated up and is relatively steady; by afternoon the seeing is always very bad indeed. Conversely, the night observer usually finds that conditions improve as the air cools down. The air may be very unsteady at 10 pm, but by the early hours it has often settled down, and those idyllic moments when a star or planet lies perfectly calm in the field usually occur not long before dawn.

The Moon is the first nocturnal object to attract attention, and to the casual observer it is of never-ending interest. Lighting conditions are always varying, and a crater or a mountain chain changes amazingly in character from night to night. A small reflector or refractor will show a maze of fine detail, and with a magnification of × 150 or × 200 a crater such as Copernicus is an unforgettable sight when caught on the terminator; it looms up, filled with shadow, like the rim of some celestial cauldron. Another magnificent scene is the Mare Crisium two days after Full, when its mountain border stands out in stark grandeur against the evening shadows. The regular lunar observer soon compiles a mental list of notable sights at different phases.

A small-scale map is essential, and the most easily available (and dependable) one is that drawn by T. G. Elger, published by George Philip & Son. This makes crater identification an easy matter, and one soon gets to know the best time at which to view different formations. If a detailed study is being made of one particular feature, however, it is important to view it under all conditions of lighting, and not simply at sunrise or sunset, when the view is naturally the most spectacular. A very interesting region is the part close to the limb, where libration squeezes craters into view at certain times and then snatches them away from our gaze.

Of course it is not possible to become an observer overnight, for the eye requires a good deal of training before it works in full co-operation with the telescope. This is particularly noticeable in the case of a small planet such as Mars. A trained observer may see considerable detail, but a casual viewer will probably have difficulty even in picking out the polar cap. Faint companions to double stars are another severe test of experience. However, there is no short cut; the only way is to keep at it, and as the nights pass so more and more markings will appear until it seems astonishing that the disk should ever have appeared featureless!

The role of the lunar observer has been drastically reduced in recent years. Cartography as such disappeared with the remarkable series of *Orbiter* photographs, which cover almost the entire surface in fine detail, and drawing lunar formations cannot now seriously be considered of scientific value. Nevertheless, such an exercise is excellent training for the eye and hand, and some superb results can be achieved if the observer is a talented draughtsman.

There are several 'notorious' areas on the Moon's surface where observers have, from time to time, reported strange obscuration of detail, and these are well worth watching; though it must be remembered that the chances of detecting anything unusual with a small telescope are rather small. Perhaps the most famous is the floor of the 60-mile crater Plato, on the northern shore of the Mare Imbrium. Another apparently active region is a part of the Mare Crisium, while Alphonsus, after its 1958 publicity stunt, has been closely watched for renewed activity. The recent interest in transient lunar phenomena, or TLP's, has been mentioned earlier, and some active members of the BAA Lunar Section are running their own patrols of likely regions. This sort of work is attractive, and can be undertaken with an aperture of 6 inches or larger provided the observer is prepared to persevere. It has been claimed that these reddish glows are more easily seen if red and blue filters are superimposed alternately just in front of the eyepiece; a coloured region will appear brighter in the light of its own colour, and appears to 'blink' when the filters are rotated rapidly. There is still a good deal of controversy over the whole subject of lunar change, and many observers have never seen anything unusual in a lifetime's study. On the other hand, others have.

Occultations, which are predicted in the BAA's annual *Handbook*, are well worth observing for their own sake, but if they are timed the accuracy must be to within a second, and preferably less. Lunar eclipses are also extremely interesting, because the sudden wave of cold which sweeps across the surface has been reported to cause temporary alteration of surface tone inside certain craters, such as Atlas and Hercules. Some are much brighter than others, and it is interesting to compare the dominant colours of different eclipses. The next total lunar eclipse visible from Britain will occur on November 29th, 1974.

Mercury will show an obvious disk in a 3-inch, and detail can be glimpsed, but serious work is out of the question except with a large observatory instrument; the main fascination is in glimpsing it at all. Venus is a much more suitable object, for it is very easily visible and shows a large disk. Its great brilliancy is actually a considerable disadvantage, because when seen at night its glare masks the detail and produces false effects. The only way to solve the difficulty is to observe Venus in the bright dawn or sunset sky, according to the elongation.

Finding a planet by daylight may sound a formidable task,

but Venus is so bright that it can, under first-class conditions, be seen with the naked eye. If its approximate place is known (and it can be worked out easily using the positions published in the BAA *Handbook*), this region of the sky is swept over slowly, using the lowest possible power to provide a wide field of view. Once the planet is found, the magnification is stepped up. Unfortunately seeing conditions are often bad, especially in the early evening, and this means that powers higher than × 100 may be less effective than lower magnifications. It is essential to have a sharply-defined disk, and if this is not possible with even a low power it is clearly hopeless to observe at all. All things considered, the Planet of Love is a very frustrating object.

There are several interesting phenomena to look for while Venus passes through an elongation. For obvious reasons the evening apparitions are likely to be better observed (it takes a good deal of resolution to exchange a warm bed for the chill of a winter dawn), and these begin when the planet is a shrunken disk just past superior conjunction. As it swings out to maximum elongation the phase lessens and the disk slowly expands, the dusky shadings, if any, becoming visible with more certainty. Because of the large variations in the diameters of both Venus and Mars, some observers adjust the scale of their drawings accordingly. The normal practice is however to keep to a standard size, a diameter of 2 inches being suitable. As a general rule, always make drawings too large rather than too small.

The Venusian markings are usually so obscure that it is very hard to make a representational sketch. One good way is to use a very soft pencil, smudging the marks with a finger and adjusting their contours with a putty rubber.

Track should be kept of the planet's progress towards the perfect half-moon phase, or dichotomy. Because of atmospheric effects, Venus always reaches this point several days earlier than predicted during an evening elongation (it is late when west of the Sun), and it is interesting to note the date and check the error, which is not always the same. Dichotomy cannot, of course, be defined precisely; the planet appears to linger for three or four days as a perfect half, and the exact date must be an average. At this stage of the apparition the weather is almost certain to be consistently cloudy![1]

In the crescent form the planet's expansion becomes much more accelerated, and by the time the phase has shrunk to 25 per cent watch should be maintained for the Ashen Light. Precautions are necessary here, for it is fatally easy to imagine seeing the dark side; the only safe way is to use an occulting bar. This can be made very simply by fastening a strip of paper across the eyepiece so that it blocks out half of the field of view and appears exactly in focus. The telescope is then manipulated so that the crescent is hidden behind the very edge of the strip. If the dark side remains visible the effect is manifestly real; if it is an illusion, it will vanish. No sightings should be recorded unless they are made using an occulting bar.

At this stage the bright 'polar caps' often become very distinct, and they should be observed carefully. Finally, there is the atmospheric extension of the horns of the crescent when Venus is very near inferior conjunction. Sometimes it is so pronounced as to form a thin diffuse circle entirely enclosing the night side, although such observations are made difficult by the Sun's proximity. It is perfectly possible to follow the planet right through inferior conjunction, when it is only a few degrees away from the blazing solar disk. This has been done with a 3-inch on several occasions, and forms an interesting challenge.

Turning to Mars, we find a less convenient state of affairs. It can be well observed only for a couple of months on either side of opposition, and these occur at intervals of two years; moreover, we are now passing through the unfavourable aphelic stage. But even at a near approach little valuable work can be done with a refractor of less than 5 or 6 inches aperture, or anything smaller than a 10-inch reflector. Mars is really one of the hardest planets to observe properly. At the time of writing, a small telescope will show the polar cap with its dark band, and the Syrtis Major and other prominent features, but

[1] Something known to all amateurs as Spode's Law.

serious observation is out of the question. Nevertheless it is fascinating to gaze at the warm, reddish-ochre disk with its tiny throne of ice and reflect that here, at least, life may be possible. Mars, tiny though it is, is the most openly absorbing of all the planets.

Some of the brighter minor planets, such as Ceres and Vesta, are well worth looking up just for the satisfaction of seeing them. The BAA *Handbook* publishes details of their positions, and from these their places can be plotted on a star chart. But in order to identify the planet itself it is necessary to do what the celestial police did; draw the field, and re-compare it with the view on the following night. The minor planet will have betrayed its nature by shifting position, and once it is found it can be followed until absorbed by the evening twilight. Hunting (and finding) minor planets is a satisfying task.

There is no real difficulty in finding Jupiter. It is cold and remote, but so large that it normally outshines Mars and shows the beginnings of a disk in binoculars. With a power of × 30 belts begin to appear, and a 3-inch refractor working with a magnification of × 150 or more will show a great amount of detail in the turbulent cloud belts.

Jupiter spins so rapidly that a period of 5 minutes will reveal a slight drift from right to left, and herein lies the main task of the Jupiter observer: the taking of 'transits'. A feature transits when it lies on the central meridian; the time is noted, and after the observing session its longitude is calculated from tables in the *Handbook*. If it survives for several Jovian days its rotation period can be worked out to within a few seconds. This work is slowly leading to more complete knowledge of Jupiter's many individual currents, and is largely in the hands of amateur observers. Jupiter, in fact, is the amateur's planet.

By comparison with transit observations, disk drawings are of limited value – they are, however, useful as a guide to the various features. A lookout should be kept for any notable colour. Vivid hues never occur, but at the present time the Great Red Spot is a greyish-pink colour, while the equatorial zone is the colour of rather strong coffee. In the past Jupiter was much more colourful than it is today, and there may be a long cycle of activity.

Mars presents itself for examination only at fleeting intervals, but Jupiter is positively brazen; it is lost near conjuction for less than three months in the year, and never appears inconveniently small, so that an almost continuous record can be maintained. Satellite phenomena are also interesting. When they pass in front of the disk they often cast a shadow on the clouds (the equivalent of a total solar eclipse), and these can be easily seen with a small telescope. Sometimes one moon occults another, and their combinations of position on both sides of the planet are endless. Jupiter's ever-changing retinue is one of the delights of the sky.

Sober Saturn, the slowest-moving of the naked-eye planets, is too famous for its beauty to need extolling. Its rings and globe are well seen, though it is rather hard to make out individual details; a 3-inch will show Cassini's Division and the equatorial belts, but that is about all. At the moment things are made more awkward by its low altitude in Britain, and it will be some years before it crosses the celestial equator and appears reasonably high in the sky. The bright moon Titan is very easy to find, and Rhea and Iapetus (at western elongation) can also be picked up without any difficulty. Saturn's considerable axial tilt means that the satellites do not appear, as Jupiter's do, strung out in an orderly line – except when we pass through the plane of its equator. This last happened in 1966, when the edge-on rings disappeared from view in small telescopes for several days.

The outer planets are much less interesting, though once again it is satisfying to locate their positions. Uranus, shining with its curious pale blue light (quite unlike that of a star), shows an obvious though minute disk; at the moment it lies in Virgo and is easily swept up. Neptune, in Libra, also offers few difficulties because of its blue-green tint. Only Pluto is too faint for small apertures.

Telescopes possess no advantage over the naked eye for observing meteors, but comets are a different matter altogether, and it is rare for a year to pass without one being visible in a

3-inch refractor. Predictions for known comets are issued in the BAA *Handbook*, while new discoveries are announced in their Circulars (they are also often mentioned in the national press, but would-be searchers are advised not to take these reports too literally). Because comets are such diffuse objects it is essential to use a low power, so as to embrace the widest possible field of view. It usually happens that comets appear brightest when situated just beyond the twilight arc, either at dawn or sunset, and because of this it is clearly desirable to have good east and west views if their observation is to become a regular part of the observatory programme. In general it is useless to observe until the sky is really dark, for the slightest light will make even a bright comet obscure. If it is very near the Sun, of course, this cannot be helped.

Comet hunting, as opposed to following known objects, is a field of work calling for immense patience; regular observers, of which there have been only about a dozen in the last two centuries, take on average about 300 hours of actual sweeping to find one comet. Perhaps this explains why so few people have taken it up! The method is simple. A region of the sky (preferably near the Sun) is selected and carefully covered by swinging the telescope across it in horizontal sweeps; at the end of one sweep it is raised or lowered slightly and another sweep taken in the opposite direction, overlapping the first. If any suspicious diffuse object is seen it is looked up on a star map, for the chances are it is a known nebula. If it is not marked, it must be watched carefully for movement; a slight shift over an hour will betray its cometary nature. If the comet is very near the Sun it will probably have grown a tail, and be identifiable anyway.

All would-be comet hunters must take as their model George Alcock, of Peterborough, an amateur astronomer who by profession is a schoolteacher. After several years of patient searching he created a sensation by discovering, in 1959, two comets within a week, and he added still further to his already considerable reputation by discovering two more in 1963 and 1965. His intense devotion to his work manifests itself in the following account.

From 1931 to 1952 I carried out naked-eye observations of meteors at Peterborough. . . . The importance of that kind of work diminished after about 1949, when it was largely superseded by radar and improved photography. So I looked around for other astronomical work and, in 1953, started on my programme of comet-seeking and searching for novae. . . .

Although I discovered no nova, and no comet until August this year (1959), my results were not altogether negative. I reported all the telescopic meteors I saw, and I managed to learn the patterns of some 20,000 stars in Galactic fields to help in detecting novae.

The comet searcher can appreciate the beauty of the night sky more than any other observer.

All this time, I felt that my 4-inch refractor was inadequate for comet-sweeping, so in 1955 I bought a 45-mm × 12 binocular and in 1957 a 100-mm × 25 telescope-binocular in very poor condition. The second of these instruments had no stand and, for the 154 hours that it was in use, was rocked on old coats on the tops of two Stevenson meteorological screens.

At last, in April this year, I bought a good 105-mm × 25 instrument. I was not able to use it much until July, and then, at the end of August, I made my two discoveries.

On August 25th, on my 560th night, and in my 646th hour of observation, I found my first comet in the northern part of Corona Borealis – a rather disappointing object, faint, and very diffuse. As it had no tail, I waited 24 hours to confirm its motion. Three comets announced between 1953 and 1959 had proved to be false ones, and I did not want mine to be a fourth. That day was a very long one indeed.

Perhaps some of you will wonder why I did not stop searching after the discovery of a comet. The answer is that I cannot sleep when I know the sky is clear.

On the night of Saturday, August 29th, the sky clouded over and I went to bed; but I woke up at 2.30 am to find the sky clearing rapidly. After an hour and a half of fruitless searching I turned to the low eastern dawnlit sky, and after a few moments picked up my second comet. As it had a faint tail I rang Herstmonceux to ask them urgently to take a photograph . . . the photograph was taken, but I was not told and so had another long day of suspense. I had not been greatly excited by the first discovery, but this was far different.

I could hardly expect an English sky to be clear again the next morning, and cloudy patches did indeed develop over a superb sky before the crescent of the Old Moon appeared. I was then afraid the comet might be occulted, but at last I saw it. The comet and the Moon were together in the field – a pretty spectacle. . . .

As the comets vanished so quickly, I have begun to doubt whether I really discovered them. It is good to know that they were really seen by others.

To anyone wishing to take up the pursuit, it remains only to wish them the very best of luck!

CHAPTER 30

Amateur Astronomy – the Stars

THE SOLAR system offers an enthusiastic amateur plenty of scope for research, but the remote, lonely stars are much less co-operative. Their study is so advanced that for most purposes specialized equipment is necessary; spectroscopic work, study of proper motions, and investigations into distant galaxies are of course far beyond the amateur's range. The only real niche is variable star work, while the occasional nova can be tracked as it dims after its blaze of defiance, but in general the depths of space must be searched strictly for their varied showpieces: double stars, clusters, nebulae, and some of the nearest galaxies. Many people dismiss this as worthless, and in the scientific sense it is; the likelihood of detecting anything unusual is negligible. But the person who does not occasionally succumb to the sheer beauty of the night sky is no astronomer, and a sight of the Double Cluster in Perseus, or the Orion Nebula, gives a feeling of awe that is all too sadly lacking in this severely technological age. Armed with a star chart and a list of objects, no clear night in the year need be without its spectacles.

Many books publish lists of the 'finest objects' visible with a small telescope, but unfortunately few seem to be based on first-hand experience, and are simply lifted from earlier compilations. This is a great pity, since it suggests that they are established and cannot be revised further, and this is very far from the case. Moreover, they include small-scale star charts that are of very limited value, since they cannot possibly reproduce the sky as accurately and clearly as a larger atlas such as *Norton's*.

The northern objects included in the following list are therefore based on the independent but joint work of two amateur astronomers, living in latitude 52½°N: John Larard (3-inch refractor) and the author (3½-inch refractor). This means, of course, that far southern objects have had to be chosen at the word of other observers. It is hoped that perusal of some of the double stars, clusters, nebulae, and galaxies included here will prompt readers to search for the many other stellar show-pieces that have had to be overlooked by the demands of selection.[1] They are arranged in alphabetical order of constellation, and are all marked in *Norton's Star Atlas*. It will be

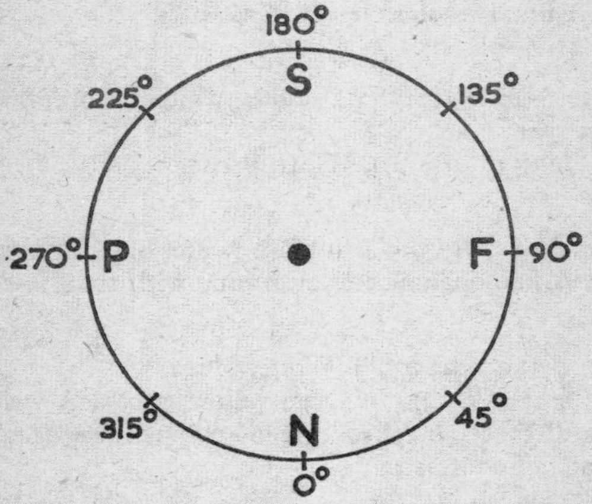

FIG. 60. *Position angle*. This shows the telescopic view, with north at the bottom. Because the directions 'east' and 'west' can sometimes be misleading, especially in the case of circumpolar stars, they are replaced by 'following' and 'preceding', terms which refer to the motion of the star through the field of view due to the Earth's rotation. There is therefore no ambiguity.

noticed that there are several disagreements over star colours; this is nothing unusual, and should act as an extra incentive for others to check up.

By each double star is given the magnitudes of the components, the distance between them (in seconds of arc), and the 'position angle' (PA), of the companion relative to the primary, which reveals in which quadrant to look. The PA dial, as seen in an inverting telescope, is shown in Fig. 60.

[1] *Norton's Star Atlas* lists over 400 double stars, as well as a great many galaxies, nebulae and clusters.

Andromeda

γ (3·0, 5·0; 9″·7; 61°). Golden-yellow, blue. Very fine.

M.31. The Great Galaxy, visible with the naked eye. It has a curious starlike nucleus, and a 3-inch shows the two satellite galaxies.

Aquarius

ζ (4·4, 4·6; 2″·8; 270°). Pale yellow, white. Close, pretty.

M.2. A fine globular cluster 7′ across.

Aries

γ (4·2, 4·4; 8″·4; 360°). Both white or blue-white. An exquisite pair.

λ (4·7, 6·7; 37″·9; 46°). White, green.

Auriga

ω (5·0, 8·0; 5″·8; 355°). Intense white, blue. Pretty.

M.37. A fine open cluster swarming with stars.

Boötes

κ (5·1, 7·2; 13″·2; 237°). White, blue.

ε (3·0, 6·3; 2″·9; 334°). Deep yellow, green. A well-known pair. In all these so-called 'green' companions, the effect is merely one of contrast.

Cancer

ζ (5·6, 6·1; 5″·7; 82°). Orange, blue. The primary is a close double (5·6, 5·9; 1″·1; 355°), elongated with 3½-inch, × 200.

ι (4·4, 6·5; 30″·8; 307°). Yellow, blue. A fine wide pair.

M.44. 'Praesepe.' A very scattered cluster of bright stars, just visible with the naked eye.

M.67. A cloud of faint stars in a fine field.

Canes Venatici

α (3·2, 5·7; 19″·7; 228°). Pale yellow, pale blue. An easy bright pair.

M.3. A beautiful bright globular cluster. Resolved.

Canis Major

M.41. An open cluster, including a red star. Easily found, 4° south of Sirius.

Cassiopeia

η (4·0, 7·6; 9″·1; 278°). Yellow, purple. A fine pair.

H.VI.42. A pleasant cluster near φ, diameter 18′.

Centaurus

α (0·4, 1·7; 4″·1; 310°). The nearest stellar group in the sky, and a magnificent object.

ω. A noble globular cluster, diameter 30′; a blazing mass of stars.

Cepheus

ξ (4·6, 6·5; 7″·3; 279°). Cream and greyish. An attractive pair.

δ (var., 7·5; 41″; 192°). Yellow and blue, very wide and easy.

Cetus

γ (3·0, 6·8; 3″·1; 294°). Yellow, green. A rather difficult double.

Coma Berenices

M.53. A globular cluster, diameter 3′, with a brilliant centre.

Corona Borealis

ζ (4·1, 5·0; 6″·3; 304°). White, blue. A superb pair.

Crux

α (1·6, 2·1; 4″·7; 119°). A fine object.

κ. The centre of a beautiful cluster of coloured stars.

Cygnus

β (3·0, 5·3; 34″·6; 55°). Yellow, pale blue. A most glorious pair, best seen with a low power. Some observers note the companion as green.

61 (5·6, 6·3; 27″·0; 140°). Yellow, yellow. A famous binary, lying in a beautiful field.

M.39. A fine open cluster.

Delphinus

γ (4·0, 5·0; 10″·2; 268°). Yellow, blue or pale yellow. A delicate object.

Draco

η (2·1, 8·1; 5″·3; 182°). A difficult but fascinating object with obscure colours.

ψ (4·0, 5·2; 30″·6; 15°). Cream and delicate lilac.

Eridanus

32 (4·0, 6·0; 6″·8; 347°). Yellow, green. A fine double.

Gemini

α (2·0, 2·9; 1″·8; 145°). Castor. Pale yellow, white. A most magnificent object, but becoming difficult with 3-inch.

M.35. A fine cluster of bright and faint stars, diameter 50′.

H rcules

α (3·0, 6·1; 4″·6; 109°). Golden, vivid green or blue. A very beautiful double.

M.13. The Great Cluster, just visible with the naked eye. A bright, condensed mass of stars.

M.92. Another fine globular cluster, little inferior to M.13.

Lacerta

8 (6·0, 6·5; 22″·3; 186°). White, white. There are two other stars near by, forming a quadruple group.

Leo

γ (2·4, 3·5; 4″·1; 120°). Cream, greenish. A beautiful pair.

Lyra

ε. The spectacular 'double double', consisting of two pairs $3\frac{1}{2}'$ apart: 4·6, 6·3; 2″·9; 2°, and 4·9, 5·2; 2″·3; 111°. All white, they are easily divided with a 3-inch, with high power.

ζ (4·2, 5·5; 44″; 150°). Yellow, white. A fine wide pair.

M.57. The Ring Nebula. A ring of glowing gas, faint but distinct in a small telescope.

Monoceros

8 (4·0, 6·7; 13″·0; 30°). Yellow and blue. Lies in a beautiful field.

H.VII.2. A beautiful cluster surrounding the star 12.

Orion

β (0·1, 8·0; 9″·5; 202°). An easy 'test' for a 3-inch. White, bluish.

ι (3·2, 7·3; 11″·4; 142°). White and greyish. There is a nebulous glow around the star.

M.42. The Great Nebula; a wonderful object, easily visible to the naked eye, glowing with a curious sea-green tint. In its midst is the Trapezium, θ, whose four bright stars are easily seen.

Pegasus

M.15. A bright and very condensed globular cluster 6′ in diameter.

Perseus

η (4·0, 8·5; 28″·4; 301°). Yellow and blue. Lies in a fine field, of which there are many in Perseus.

H.VI.33, H.VI.34. The Double Cluster. An utterly superb object, and the finest stellar sight in the northern heavens. A very low power is required to fit both the clusters (45′ apart) in the same view.

M.34. A fine open cluster, with a double star at the centre. Diameter 50′.

Pisces

ζ (4·2, 5·3; 23″·7; 63°). White, olive; a neat pair.

α (4·0, 5·0; 2″·4; 294°). Greenish, blue. An interesting close double.

Sagitta

θ (6·0, 8·5; 11″·6; 330°). White, blue. Another star nearby makes the group triple.

Sagittarius

M.22. A very fine globular cluster, only slightly inferior to M.13.

Scorpius

β (2·0, 5·1; 13″·7; 23°). Yellow and greenish. A most beautiful pair.

α (1·0, 6·5; 3″·2; 275°). Brilliant red and green. A difficult test, for atmospheric turbulence usually conceals the companion.

M.6. A beautiful open cluster.

M.7. A very bright open cluster. Both M.6 and M.7 are unfortunately too low for observers in British latitudes.

Scutum

M.11. A fine fan-shaped cluster around a mag. 8 star.

Serpens

θ (4·0, 4·2; 22″·3; 103°). White, white. A glorious pair.

Taurus

α (0·8, 11·2; 121″; 33°). Red, blue. Not spectacular, but a very useful light test. Under good conditions a 3-inch should show the companion.

M.1. The Crab Nebula, a small dim object near ζ.

Two well-known open clusters, the Hyades and the Pleiades, are to be found in Taurus.

Triangulum

ι (5·0, 6·4; 3″·9; 74°). Yellow and blue. Very fine.

Tucana

47. A magnificent globular cluster, visible with the naked eye. Second only to ω Centauri.

Ursa Major

ζ (2·1, 4·2; 14″·5; 150°). White, pale green. The most famous double in the sky, and a fine object.

M.81. The bright nucleus of a spiral galaxy.

M.82. Another spiral, appearing as a curved nebulous ray $7' \times 1\frac{1}{2}'$. This is because we are seeing it edge-on. It is very close to M.81.

Ursa Minor

α (2·1, 9·0; 18″·3; 217°). The Pole Star. Yellow, bluish. An easy object with a 3-inch.

Virgo

γ (3·6, 3·7; 5″·7; 317°). Both cream or white. A superb pair, one of the finest in the northern sky.

Vulpecula

M.27. The Dumb-bell Nebula. Two hazy patches of light, best seen with a very low power.

The magnification necessary to show these different objects well varies tremendously. All nebulae and galaxies require a very low power to preserve their misty contours, while open clusters, which are scattered over a relatively large area of sky, also demand a large field of view. Globulars, on the other hand, are small and condensed, and require a high magnification if any of their individual stars are to be made out. A powerful eyepiece destroys the contrast of a wide double such as β Cygni by spreading the components too far apart, but ζ Aquarii and ε Boötis can be divided only under a high power. Because of this, close double stars are an excellent test of the telescope's defining capacity, and it is possible to choose certain doubles as test objects for different apertures. For instance, a 3-inch refractor should be able to divide the star ε Arietis (5·7, 6·0; 1″·5; 205°), if conditions are first-class, since the distance between the components is equal to the telescope's limit of resolution. On the other hand, ε Arietis would be no test for a 12-inch telescope, whose user would have to find a double whose components were only 0″·4 apart.

There are other idiosyncrasies also. If one member of a pair is much brighter than the other, it tends to obscure its companion; α Scorpii (Antares) is a case in point, and so of course is Sirius. In fact, every double demands its own special conditions, and this is partly what makes the subject so fascinating. There is another, less encouraging, aspect. Professional astronomers, while keeping watch on close binary pairs (which are mostly of little interest to the amateur), have neglected the wider and more spectacular doubles, many of which are optical. Since their separations progressively change with time, due to the proper motions of their components, measures become out of date. Any amateur with the time and equipment would be doing a great service to his fellows by measuring some of these bright pairs.

The main realm of the amateur in stellar astronomy is the observation of long-period variable stars. Some of these have already been mentioned as naked-eye objects, but there are literally hundreds of fainter ones, usually with periods of about a year. The BAA Variable Star Section has about 70 variables on its programme, although some of these can be followed to minimum only with large apertures. The method of estimation is essentially similar to that employed with naked-eye stars, except that the comparison stars must lie within the same field as the variable itself, to allow a direct estimate. The Section issues charts to its members stating the magnitudes of these comparison stars. As well as observing exclusively telescopic variables, it is interesting to watch a star such as Mira slowly brighten until it is visible to the unaided eye. *Norton's* lists a great many of these objects, and anyone interested will certainly find more than enough material to keep himself occupied.

Variable stars are in general designated by one or more capital letters, such as R Geminorum, SS Cygni, and so on, and anyone who believes that their observation is always a leisurely affair would do well to pick up CY Aquarii, an RR Lyrae star with a period of only 88 minutes. At one point in this cycle its magnitude leaps by 0·7 in 9 minutes, which gives the observer the spectacular privilege of literally seeing a star brighten before his eyes! Unfortunately it is rather faint, and is most easily seen with apertures in excess of about 8 inches, but under good conditions a 3-inch refractor should be able to show it. The range of magnitude is from 10·5 to 11·3.

Sweeping for novae, another potentially valuable occupation, requires the same phenomenal patience as comet hunting; also, wide-angle binoculars of the type Alcock uses are more suitable than a small refractor. But it is a good idea, as part of the night's routine, to scan the constellations for an unexpected star. The chances are never very high, but on the other hand they are not negligible; after searching the constellations for doubles the star patterns soon become familiar, and the likelihood of noticing a new star is very high indeed. The most fruitful region is in the track of the Milky Way, for here we are seeing more stars per square degree than anywhere else, and nearly all the naked-eye novae have occurred in or near its depths. When a nova does occur it is interesting to watch it fade, for momentary brightening sometimes interrupts its progress, and there may also be perceptible colour change; Nova Herculis 1934 acquired a greenish hue late in its career. Indeed the amateur's motto, whether observing the Sun, planets, or variable stars, should be never to take anything on trust. It is by persistently checking up, and eventually discovering unexpected vagaries, that the greatest discoveries are made.

Celestial photography is practised much less than it should be. It naturally requires an equatorial telescope, but in occasional emergencies, such as the appearance of a bright comet, perfectly satisfactory mountings have been improvised from remarkably un-astronomical equipment; one observer obtained some excellent photographs of Comet Mrkos, in 1957, by using a camera constructed from a 4-inch portrait lens, a dog's kennel, a fence paling (discarded!), a great deal of whiskery Post Office string, and two empty cornflakes boxes. The same observer, incidentally, discovered a bright comet on Boxing Day, 1960, when testing a small telescope! Less primitive apparatus is in some ways desirable, however, and anyone interested in these and other matters to do with instruments would do well to consult *Amateur Telescope Making*, Vol. 1, published by the Scientific American. Some amateurs are experimenting with colour photography, and their results are naturally of the greatest interest.

All in all, the amateur astronomer of today has little excuse to be idle; he may never make a great discovery, but the heavens contain more than he can hope to see in a lifetime. And if the greater part of this book has been about the problems posed by modern astronomical investigation – problems that are far beyond the range of the non-professional – we must never forget that the basic purpose of any science is to stimulate the mind. A person is no more answering the call of the night sky by reading a book about it than he is journeying to a foreign country by reading a travelogue. Both may describe sights that he can never hope to witness, but without the element of participation he is no traveller at all. It is therefore fitting to end with the plea not to be satisfied with words and pictures, but to buy a telescope and look, however inadequately, at the universe face to face.

APPENDICES

APPENDIX I

Some Astronomical Terms

(The definitions given here do not pretend to be exhaustive, but they should be complete enough for most purposes)

Aberration. The apparent displacement of a star caused by the 'bending' of its light, due to the Earth's orbital motion. Its effect is to make every star appear to revolve around a fixed point, its maximum distance from this point being about 20″·5. *Chromatic aberration.* The formation of a coloured fringe around the image formed by a simple lens.

Albedo. The ratio of light reflected to that received. Venus, for example, reflects 59 per cent of the sunlight falling on it; its albedo is therefore 59 per cent.

Annular eclipse. A solar eclipse occurring with the Moon near apogee, so that it cannot cover the Sun completely.

Aphelion. The point on a planet's or comet's orbit which is at the greatest distance from the Sun.

Apogee. The point on a satellite's orbit-which is at the greatest distance from its primary.

Appulse. An apparent close approach of one celestial body to another. It is purely a line-of-sight effect, and does not imply physical proximity.

Asteroid. An alternative name for a minor planet.

Astronomical Unit. The mean distance from the Earth to the Sun, now taken as 92,900,000 miles.

Baily's beads. Phenomenon occurring at the beginning and/or end of a total solar eclipse, when fragments of the photosphere shine out brilliantly through deep rifts in the Moon's limb.

Barycentre. The point on the imaginary line joining two mutually-connected bodies around which they revolve.

Binary system. Two stars revolving around each other under mutual gravitational attraction.

Celestial sphere. An imaginary sphere carrying all celestial objects, rotating in 23 hours 56 minutes, and inscribed with the celestial equivalents of poles, equator, latitude, and longitude.

Circumpolar. An object so close to the celestial pole that it remains permanently above the horizon.

Conjunction. Strictly speaking, the condition of two celestial bodies when their RA or Dec. become the same. In practice it is used as a synonym for appulse. For instance, a superior planet is in conjunction when it appears near the Sun in the sky. *Inferior* and *Superior Conjunction.* An inferior planet's appulse to the Sun on the near side and the far side of its orbit.

Constellation. A defined region of the celestial sphere.

Culmination. The condition of a celestial object when at its highest possible altitude above the horizon. Unless very near the pole, this occurs when it is due south.

Cusp. A horn of the crescent Moon, Mercury, or Venus.

Declination. The angular distance of a celestial object north (+) or south (—) of the celestial equator.

Doppler effect. The shift of a source's spectral lines due to its motion towards or away from the observer.

Earthshine. Illumination of the Moon's dark side due to sunlight reflected back from the Earth.

Eclipse. Passage of the Moon wholly or partly across the Sun, or the passage of a satellite wholly or partly through its primary's shadow.

Ecliptic. Apparent path of the Sun around the celestial sphere, marking approximately the plane of the solar system.

Elongation. The position of an inferior planet at its greatest angular distance from the Sun.

Epoch. Generally speaking, the date for which star positions (as on a chart) are correct.

Equinoxes. The two points at which the ecliptic crosses the celestial equator. These lead to the fearsome term *Equinoctial Colure*, which is the great circle on the celestial sphere passing through the poles and the equinoxes.

Gibbous. Phase intermediate between Half and Full.

Hour Angle. The interval (measured in sidereal time) since a certain celestial object was last on the meridian.

Inferior planet. A planet whose orbit is smaller than the Earth's.

Julian Date. The number of days that have elapsed between the day in question and January 1st, 4713 BC. It is expressed in days and decimals of a day, and is used in much computing work. The selection of the original date is purely arbitrary.

Libration. The axial swinging of the Moon with respect to the Earth, or of Mercury with respect to the Sun.

Limb. The edge of the apparent disk of a celestial body.

Magnitude. Classification of a star's real or apparent brightness. Hence *Absolute magnitude* and *Apparent magnitude.*

Meridian. The great circle passing through the zenith and touching the horizon at the north and south points. The meridian of a planet or the Moon is the line joining the north and south poles and passing across the centre of the disk.

Nadir. The point on the celestial sphere directly beneath the observer.

Node. The apparent crossing of two paths or orbits, such as the celestial equator and the ecliptic at the equinoxes.

Nutation. A minute oscillation of the Earth's axis, due to lunar perturbations, superimposed on the much more marked precession.

Occultation. The passage of a nearby celestial body in front of a more remote one.

Opposition. The state of the Moon or a planet when opposite the Sun in the sky.

Parallax. The apparent displacement of a body against its background when seen from different stations.

Perigee. The point on a satellite's orbit which is closest to its primary.

Perihelion. The point on a planet's or comet's orbit which is closest to the Sun.

Period. The time taken for a planet, comet, or satellite to achieve one circuit of its orbit. *Sidereal period:* relative to a fixed point. *Synodic period:* relative to the Sun.

Personal equation. The fractional discrepancy between the observation of a phenomenon (such as an occultation) and the recording of it.

Perturbation. The influence of a celestial body on the orbit of another nearby body.

Position angle. The bearing of the fainter member of a double star measured from its primary. It is reckoned in degrees, starting at the north point and working counter-clockwise.

Precession. A slow 'wobble' of the Earth's axis that takes 25,900 years to complete. This has the effect of constantly changing the celestial co-ordinates.

Retrograde motion. Real or apparent motion of planet, comet, or satellite in the opposite sense to that usual in the solar system. It can also be applied to binary stars.

Right Ascension. The celestial equivalent of longitude, measured eastward in hours from the Vernal Equinox.

Saros. An interval of roughly 18 years 10¼ days, after which the Sun and Moon are in the same relative positions in the sky. It was an ancient method of predicting eclipses.

Seeing. The state of the atmosphere, whether steady or unsteady. Some observers rate it from 1 to 10, 1 being hopelessly bad and 10 unattainably good.

Sidereal Time. Time measured on the basis of the rotation of the Earth with respect to the stars.

Solar Time. The normal civil time, being measured on the basis of the Earth's rotation with respect to the Sun. There are two kinds: Apparent Solar Time, as measured by a sundial, and Mean Solar Time, which smooths out the Sun's irregular motion along the ecliptic, in turn brought about by the eccentricity of the Earth's orbit.

Solstices. The two points on the ecliptic farthest removed from the celestial equator; the Sun is in these points at midsummer and midwinter. Hence *Solstitial Colure*, the great circle passing through the solstices and the celestial poles.

Southing. A synonym for culminating.

Superior planet. A planet whose orbit is larger than the Earth's.

Transit. There are three meanings of the word. A star or planet transits when it crosses the meridian; a detail on a planet's disk transits when it is carried across the planet's meridian; and a satellite transits when it crosses in front of the disk of its primary.

Vertex. That point on the limb of the Moon or a planet which is highest above the horizon.

Zenith. The point on the celestial sphere directly above the observer.

Zodiac. The zone 18° wide, centred along the ecliptic, inside which the major planets except Pluto are always to be found. The constellations through which it passes are called the Zodiacal constellations.

APPENDIX II

The Greek Alphabet

α Alpha
β Beta
γ Gamma
δ Delta
ε Epsilon
ζ Zeta
η Eta
θ Theta
ι Iota
κ Kappa
λ Lambda
μ Mu
ν Nu
ξ Xi
ο Omicron
π Pi
ρ Rho
σ Sigma
τ Tau
υ Upsilon
φ Phi
χ Chi
ψ Psi
ω Omega

APPENDIX III

Joining a Society

BY THIS time the reader will have been able to decide whether or not astronomy holds any fascination for him. If it does, the next step is clearly to join a society, and the leading one in Britain is the British Astronomical Association, which holds monthly meetings in London (where it has an extensive library). The BAA is an amateur body, and full details can be obtained from the Assistant Secretary, The British Astronomical Association, Burlington House, Piccadilly, London W.1.

The BAA has branches in Australia, but American amateurs are served mainly by two observational societies: the Association of Lunar and Planetary Observers (inquiries to Box 26, University Park, New Mexico), and the American Association of Variable Star Observers, at Cambridge, Massachusetts. Many other countries also have their own amateur groups.

Mention should also be made of an organization designed to help young newcomers to the science: the Junior Astronomical Society (Secretarial address: 96 Elmbourne Road, London S.W.17). Active observers may also be interested in *The Astronomer*, a monthly magazine devoted to observational reports. Details can be obtained from the Secretary at 130 Derinton Road, London S.W.17.

INDEX